FROM SCAPA TO JUTLAND

THE STORY OF HMS *CAROLINE* AT WAR FROM 1914–1917

John Allison

To an excellent mentor the late Captain A.S.P. Orr
and all who have served in *Caroline*

Published 2019 by Colourpoint Books
an imprint of Colourpoint Creative Ltd
Colourpoint House, Jubilee Business Park
21 Jubilee Road, Newtownards, BT23 4YH
Tel: 028 9182 6339
Fax: 028 9182 1900
E-mail: sales@colourpoint.co.uk
Web: www.colourpoint.co.uk

First Edition
First Impression

A catalogue record for this book is available from the British Library.

Designed by April Sky Design, Newtownards
Tel: 028 9182 7195
Web: www.aprilsky.co.uk

Printed by GPS Colour Graphics Ltd, Belfast

ISBN 978-1-78073-124-7

Front cover: HMS *Caroline* undergoing full power sea trials off Birkenhead, December 1914.
Courtesy Wirral Archives Service

CONTENTS

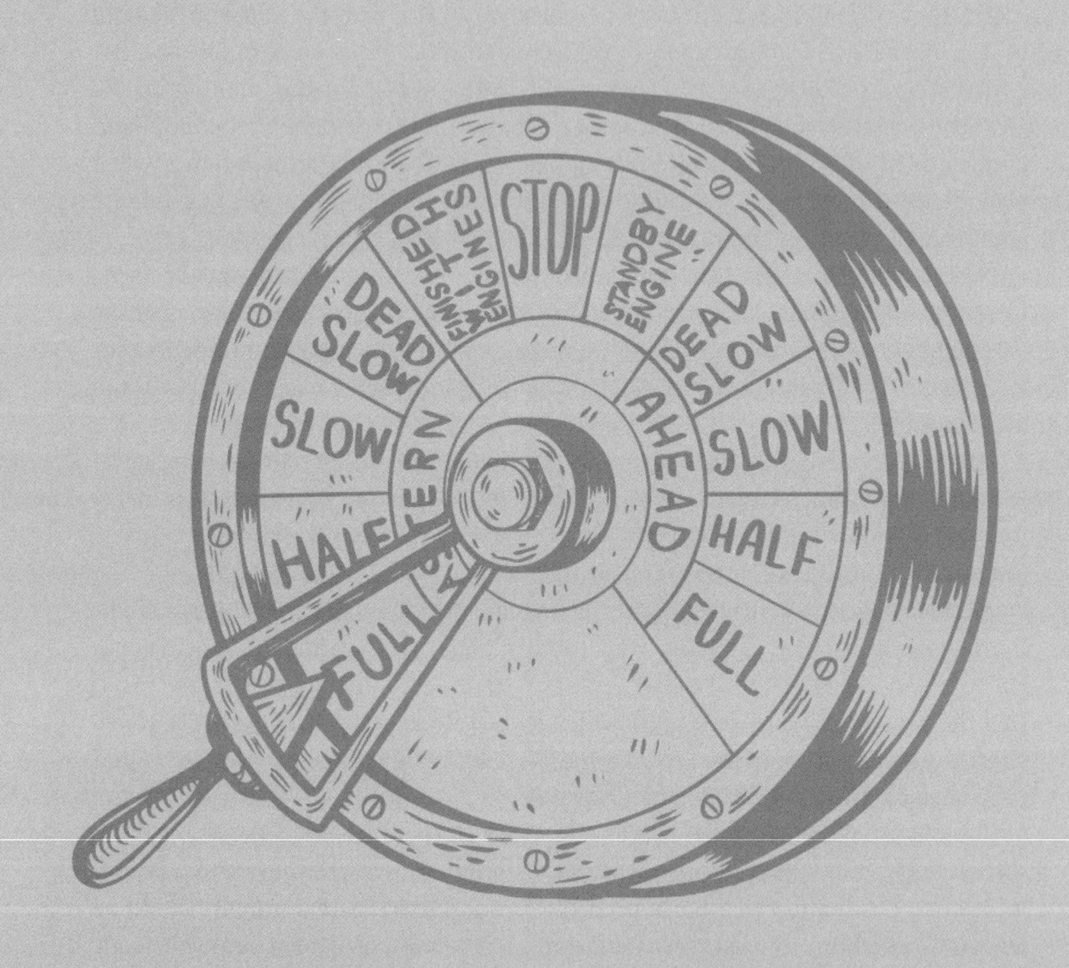
DEAD SLOW
FINISHED WITH ENGINES
STOP
STANDBY ENGINE
DEAD SLOW
SLOW
HALF
FULL
AHEAD
SLOW
HALF
FULL

INTRODUCTION

AMID THE TWISTS and turns of her survival to this day, the story of the light cruiser HMS *Caroline* spans a century and more. This book focuses on her early career, the role she played as just one of many components making up the Grand Fleet in time of war. We look at her routine participation in contraband control and, most dramatically, her appearance at the Battle of Jutland, when providence smiled upon her and guaranteed a safe emergence from that intense cauldron of explosion and fire.

As one of His Majesty's latest and most technologically advanced light cruisers, *Caroline*, was complete by early December 1914. King George V, we may surmise, was delighted at this new arrival, for he was Great Britain's modern-minded 'sailor king', being keenly interested in the fighting navy and having had a career at sea to look back on.

The enemy also had intelligence of *Caroline*.[1] The Imperial German naval mastermind, Grand Admiral Alfred von Tirpitz, knew too that *Caroline's* direct forbear, *Arethusa*, had wreaked havoc amongst a number of German light cruisers only three months ago in the seas near the fortified island of Heligoland.

So, to avenge this setback, once *Caroline* made her debut in the North Sea, high-flying reconnaissance Zeppelins[2] were to sally from their gigantic hangers in north Germany to locate her, to spot her with powerful Zeiss binoculars with the result that the new English cruiser might be dispatched straight to the bottom, *Caroline's* short-lived career terminated. And what a great public relations coup that would have been for the Imperial German Navy!

First however, power trials had to take place in the Firth of Clyde. Those trials satisfactorily done, it would be a case of shaping a northerly course, to a wilder location, a destination that harboured the Grand Fleet. There a warship could be tested, 'worked up', in Navy vernacular, trained at sea to a state of peak efficiency amid glorious scenery on a fine sunny day – but also in a place where the vilest storms could be unleashed on a bad day. Yes, her destination was none other than Scapa Flow. A bleak destination then, of which more when we meet the captain's steward, the meticulous keeper of a private diary.

Thus, *Caroline* and her crew were to set out on a great undertaking, one not without grave risks but certainly one that held forth a measure or prospect of glory.

1 Details are included in the Imperial German Navy annual; *Nauticus*, 1914
2 Manned airships

My father Dr RS Allison's account of the history of HMS *Caroline,* published in 1974 to mark her sixtieth anniversary, expressed the hope that, "we may be forgiven if we begin to feel that *Caroline* is one of the immortals… she belongs to that company of great warships of the past which are preserved in the ports of the world: *Victory, Unicorn, Belfast* in the United Kingdom; *Prins Hendryks* in Den Helder; *Constitution* and *Texas* across the Atlantic; *Aurora* in Leningrad. They are a select and honourable company, relics of a life that is now utterly vanished."

Caroline is indeed, I believe, assured of a place in that proud company!

John Allison
Highgate, 2019

Part One

DEAD SLOW
FINISHED WITH ENGINES
STOP
STANDBY ENGINE
DEAD SLOW
SLOW
AHEAD
FULL
FULL

1

Into Her Rightful Element – The Launch

Monday 21 September 1914, 10.20am, Birkenhead

There was a war on. On the 3 August German troops had invaded Belgium, capturing Brussels and advancing onwards to Paris. Although the British and French Allies had now halted the attack, both sides were trying to outflank each other northwards towards the coast. At home, the House of Commons had decided to shelve the issue of Irish Home Rule due to the declaration of war. At sea Vice Admiral Sir David Beatty had burst upon the scene by entering the enemy's backyard, sinking three German light cruisers and two of their destroyers on the 28 August – an encounter that was already known as The Battle of the Heligoland Bight.

As First Lord of the Admiralty, the Right Hon Winston Churchill, had felt wholly justified in his personal choice (over the Sea Lords' heads) of David Beatty to command the Grand Fleet's striking force of battle cruisers, those fast, heavily gunned but somewhat lightly armoured lithe, panther-like scouts of the seas. Beatty had shown decisiveness, pluck and to Churchill's delight, the necessary offensive spirit; an attitude that smacked of the 'Nelson Touch'.

Now despite this war, work-in-progress today had come to a standstill on one of the slipways of a famous Merseyside shipbuilder, Cammell Laird, Birkenhead. The men had downed tools. It was in fervent anticipation of the launching of the vessel they had sweated upon – their traditional right. Thus, slipway number 5, Tranmere, was the focus of eager attention by riveters, fitters, shipwrights, joiners, engineers, naval architects and even some office staff. A steadily growing number of guests were arriving as well. Amongst them the Guest of Honour, Lady Lawrence Power, wife of Admiral Sir Lawrence Power, KCB, CVO, who would have a time-honoured duty to perform.

All had gathered to witness the birth of the first of a new generation of light cruiser for His Majesty's Royal Navy. However, to her builder Cammell Laird Ltd,

CAMMELL LAIRD
& CO. LTD.,
BIRKENHEAD.

DESIGNERS AND BUILDERS
OF
WARSHIPS OF ALL TYPES
FOR
BRITISH AND FOREIGN GOVERNMENTS.

LONDON OFFICE: — 3, CENTRAL BUILDINGS, — WESTMINSTER, S.W.

Advert appearing in The Shipbuilder Monthly *magazine (Vol XII No.56), April 1915.*

she was still known as Yard No 803. These onlookers took in a final vision of the long, slender, graceful hull resting on its specially constructed timber cradle. And the sharp, clipper stem, literally made an uplifting sight.

Designers and engineers of Cammell Laird Ltd, builders of warships and merchant ships, were especially interested in this particular launch. Would it turn out to everyone's entire satisfaction? Would this be a routine delivery or would there be an awkward labour?

If things went well, Cammell Laird would be able to claim an impressive first amongst the shipbuilding fraternity, for the keel plates of this vessel had been laid as recently as the 28 January – in other words, a build in a record time of 235 days.

Indeed, the embryo was, at this stage in life, well advanced. Once she entered her rightful element, the agitated, turbid waters of the River Mersey, she would displace 2,377 tons. Hull plates and angles were nearly complete. Her rudder, cross head and steering gear were in place. Nearly three tons of watertight doors, her capstan and 14 tons of plumbing, sidelights, and communication tubing were on board. Also eight tons of armament, one anchor and eight lengths of chain cable had been installed. Already fitted were 415 tons of dynamos, engines, auxiliary machinery, boilers, funnels and shafting.

Lady Power's presence was now required on the launch platform. Numerous

lumbermen stood by. The equipment had indicated a movement of $\frac{1}{8}$ inch on both port and starboard sides – before dog shores and bilge blocks had been removed. Now after the removal of further dog shores, $\frac{1}{4}$ inch of movement and, after taking away more bilge blocks, $\frac{3}{8}$ inch and $\frac{5}{16}$ inch of movement was indicated on port and starboard sides respectively.

No 803's builders had thoughtfully applied 15 hundredweight of tallow and 10 hundredweight of soft soap to her launch ways. Even though Cammell Laird could boast long and varied experience of launchings from their yard, a smooth, flawless delivery on every occasion was never guaranteed.

Classical legend has it that Athena's body in full armour sprang from the axed head of Zeus to the sound of echoing thunder on Mount Olympus. The stuff of mythology. Yet here before onlookers' eyes was no myth, rather the genesis of a legend yet to live, poised proudly and patiently awaiting her big moment.

Any minute now there would be the increasing din of fracturing stocks, crashing metal sling plates, timber blocks under great stress suddenly giving way as No 803 took her leave of Neptune's cradle to enter her element.

But time and tide would not wait on ceremony. By 11.15am all was ready for the instant she would commence her plunge down the slipway. There would have to be enough water under her keel for No 803 to float buoyantly. Her patrons had thought of this, as there was 18 ft 11 in of tide available.

Now, cometh the hour. At 11.20am precisely, the sling plates, the steel angle brackets, the wedges, the blocks, the rollers, the whole paraphernalia of the launching cradle fell away and No 803 started her plunge. And on that cue Lady Lawrence Power performed her duty that day: "I name this ship *Caroline*. May God bless her and all those who sail in her…"

Without any hesitation, or any deviation and most certainly without the need of any prompting, His Majesty's Ship *Caroline* slid with dignity down the slipway.

The measured, restrained comments of *Caroline*'s designer appended to her 'Launch Particulars' summed up a certain glow of pride on the part of Cammell Laird Ltd in their latest achievement for the war effort: "This launch was very satisfactory. There was no sign of the ship having strained herself during the launch."

Apart from the Admiralty, was it at this very moment when *Caroline* made her confident, stylish debut that another presence, albeit invisible, embraced her – the presence of 'Lady Luck'?

2

A Certain Guest

THE LAUNCH COMPLETED, *Caroline*, now the name ship of a new class of light cruiser, was towed to a nearby fitting out wharf. Rumour had it that the war would be over by Christmas. While technicians ministered urgently to her over the next weeks to prepare for her speedy arrival in the Fleet, we turn back the clock to look at the state of the Royal Navy in 1911, three years before the declaration of war. It was a pivotal year in shaping that fleet for the conflict to come.

From the 1890s to 1911 there took place many advances in the design of both capital warships and small craft, changes that greatly increased the Royal Navy's fighting power. Indeed, when King George V reviewed his Fleet at Spithead on Saturday 24 June 1911 from his yacht, HMS *Enchantress* he saw, amid the parallel lines of anchored ships, a re-formed Senior Service. Long gone were many outdated battleships of the late Victorian time. Their place had been taken by recently completed heavily armoured 'dreadnoughts', for example: *Neptune, Vanguard, Temeraire,* and *Bellerophon.* Although some older, armoured, 'protected' cruisers[1] still had a place in the lines, for example: *Warrior, Defence, Achilles, Black Prince, Good Hope*, they were up-staged by new lithe, panther-like battle cruisers: *Invincible, Indomitable, Inflexible.* This sailor king's eye also took in numerous up-to-date, fast torpedo boats and destroyers such as: *Acorn, Alarm, Rifleman* and *Larne.*

These new classes of warships were a result of the emerging threat and destructive power of two weapons: the mine and the torpedo. Also by 1911 naval guns could hurl heavy explosive shells to a range of several miles. This had acted as a spur to improvements in ship mobility. Lofty, towering piston engines were out of favour now. In their stead were space-saving horizontally aligned steam turbines. Excess machinery space allowed extra room for increased armament and armour protection.

Oil fuel for raising steam in ships' boilers was now surpassing the use of coal, a dirty, dusty fuel that took up long hours of thankless, tedious labour when a ship filled up her bunkers.

1 A type built at the turn of century constructed with early armour plating. Out of date by 1914.

DREADNOUGHTS

HMS *Dreadnought*, an overnight 'game changer', ran full power trials on 3 October 1906 after a record build time. Its armament exceeded that of any existing battleship: ten 12-inch guns mounted in five turrets, three placed along the centre line of the ship and one each on the port and starboard sides. In this way up to six guns could fire salvoes forward, six aft, and eight to either port or starboard; a uniform barrage from a standard calibre rather than one from mixed sized calibres. Range finding was made easier. The ship's spotter could make out the fall of shot better from a uniform salvo than from a salvo of different calibres. Engined with turbines, *Dreadnought* was capable of a speed of 21 knots. The hull had a belt of steel hardened armour to a thickness of 11 inches, reducing to 6 inches forward and to 4 inches at the after end.

By 1911 tactical manoeuvres at sea were being conducted at speeds up to 18 knots. Pre-dreadnought battleships were hard put to reach 14 knots when steaming at full speed.

As naval secretary to Winston Churchill, First Lord of the Admiralty, David Beatty had presciently confided to him that in any future conflict at sea the issue would be decided by each side's fast moving battle cruisers.

As if to reinforce Beatty's forecast, a particularly interesting guest warship showed up. She took her place in the line between Russia and Austria-Hungary. As HMS *Enchantress* passed down the lines of warships, thunderous salutes and the sound of 'cheering ship' echoed across the waters that sunny Saturday, 24 June. It was a day that saw dense crowds throng the sea front at Cowes and the shoreline at Southsea.

Many foreign guest warships rode peacefully at anchor. They made a majestic sight as they lay in calm repose upon the white-flecked blue waters of the Solent. Yet senior British naval officers and officials in bowler hats frowned at this particular guest with a mixture of both curiosity and concern, for her name was SMS *Von der Tann*. No mistaking it, here before their eyes lay Germany's latest battle cruiser – His Majesty's Ship, (*Seine Majestaets Schiff*). 'His Majesty' was the Emperor Wilhelm II of Germany, who was more colloquially known amongst British blue jackets as 'Kaiser Bill'. *Von der Tann* had been commissioned as recently as February 1911, the German emperor's answer to *HMS Invincible, Inflexible* and *Indomitable.*

Built by Blohm & Voss, Hamburg, *Von der Tann* was named after Ludwig von der Tann, a Bavarian general. The warship was a regular winner of German fleet gunnery prizes. She had an eye-catching profile: distinctive cranes, two squat, widely spaced funnels and armour protection superior to any current British

design. But she carried a marginally lighter punch than British battle cruisers: eight 11 inch guns in four turrets. In some thirty-five months time and but a few days before Christmas 1914 it would fall to *Von der Tann* to visit death and destruction upon an unsuspecting English east coast town.

While other world powers might have been able to afford a token dreadnought or two, Germany was openly building a dreadnought fleet with a scouting force of battle cruisers to match that of her maritime rival Great Britain. Thus it was that an arms race between Kaiser Wilhelm II's growing navy and the world's mightiest navy intensified.

HOW TO MAKE ARMOUR PLATE

The steel for armour plating a battleship was manufactured on the open-hearth system, the furnaces being heated by gas. After smelting the charcoal pig iron and mixing the alloys, which occupied about fourteen hours, the steel was run into a huge mould to form an ingot 42 inches in thickness, and weighing 56 tons.

This ingot was then taken to the 6,000-ton hydraulic forging press and reduced to a slab 24 inches in thickness, and now weighing 33 tons, its length and width being increased to the necessary dimensions.

It was then rolled down to the finished sizes, after having been re-heated for 15 hours, consuming 15 tons of coal.

This manufacturing method was used at the Sheffield Cyclops Works. By Queen Victoria's Silver Jubilee in 1897 the Cyclops Works employed more than 10,000 steel workers, such was the demand for warships for the Navy.

A contemporary pamphlet from the Cyclops Works, 21 May 1897.

3

CAROLINE'S LINEAGE

A VITAL MEANS to success in any sea battle was obviously accurate foreknowledge of enemy intentions. Dreadnoughts and battle cruisers alike depended upon information received from scouting ships: light cruisers whose job it was to both shadow the enemy's movements and report on his speed and course steered. Therefore a need arose to tailor-make a modern light cruiser, ideally suited to the role of scout but also to have the ability to lead and support a flotilla of destroyers in all sea weather states.

Various classes of outdated light cruisers were still in commission but none nimble enough to match the speeds of the latest capital ships. By 1911 the Chatham Class appeared. These were three ships: *Chatham, Southampton* and *Dublin.* Of 5,400 tons they could manage 25 knots. Their main armament consisted of eight 6 inch guns. The year 1912 saw the introduction of a variant of the Chatham Class – three ships: *Nottingham, Birmingham, Lowestoft,* mounting nine 6 inch guns and having a speed of 24.75 knots. In the illustration of *Birmingham* can be seen the shape of a familiar hull profile, particularly the bow, sharp and curving away, protruding below the waterline to suggest the presence of a ram.

Although these ships relied on coal to fire their boilers they did have the benefit of turbine drive. However, a vessel still more fleet of foot was demanded, a light cruiser combining oil fired boilers and turbines.

Arethusa emerged in early 1914. At 3,500 tons this ship had a much improved speed of 29 knots, an armament of two 6 inch guns and six 4 inch guns and mountings for surface launched torpedoes. Hull protection was provided by a belt of armour to a thickness of 3 inches, deck armour to 1 inch. Her eight boilers burned oil fuel, supplying steam to four Parsons turbines. Ships in this class included: *Arethusa* (name ship), *Aurora*, *Galatea*, *Inconstant*, *Royalist*, *Penelope*, *Phaeton* and *Undaunted.*

Clearly, by 1914 with the arrival of the Arethusa Class, a family of fast, modern light cruisers, tailor-made for reconnaissance and support operations in the North Sea, had evolved. However, by the close of 1914 the *Arethusa* specification was to

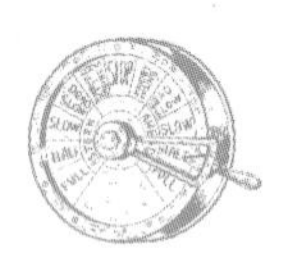

be augmented and refined yet further. Admiralty designers now offered the ultimate light cruiser – a vessel of 3,750 tons, ten feet longer overall plus one and a half feet broader in the beam, mounting eight 4 inch guns (two more than *Arethusa).* The name ship of this new class of light cruisers, as we have already seen, was HMS *Caroline.*

Advert appearing in The Shipbuilder Monthly *magazine (Vol XII No.56), April 1915.*

Appropriately, the very name *Caroline* evokes in this instance the most aristocratic of pedigrees. So far as one is able to trace, the name first came into the Navy through the re-naming of a small man-of-war launched at Sheerness in 1700. It was later converted to a Royal Yacht for George II, who had married Caroline, daughter of the Margrave of Brandenburg-Ansbach. A second Royal Yacht, built at Deptford, was launched in 1750 as *Royal Caroline.* A third vessel, said to be named after the third daughter of King George II, was the *Princess Caroline* of 80 guns. The view is taken by some experts[1] that today's HMS *Caroline* derives from a different royal personage, Princess Caroline of Brunswick, George IV's consort.

While *Caroline*'s sister ships *Carysfort, Cleopatra, Comus, Conquest* and *Cordelia,* are long extinct, the profile of their representative today certainly perpetuates and still embodies a graceful, almost regal style; the curving, sharply defined bow and low freeboard afterdeck belie a business-like purpose.

The crews of these vessels one hundred years past, were imbued with a deep sense of purpose and duty for it was wartime. In the battle line the place of these swift, elite cruisers was some miles ahead of the lumbering squadrons of heavy ships. It was an exposed position, exposed to vile weather and the first ranging shots of an equally vigilant enemy. So, even as they relied upon their high speed, they could be standing into mortal danger. They would be the first to make out smoke on the horizon, to surge forward to investigate it more closely. Friendly forces or foe? Then, if the latter and certain of the identity of the distant ships, make the age-old signal from the time of Nelson… "Enemy In Sight."

1 Such as TD Manning and CF Walker, authors of *British Warship Names* (Putnam, London, 1959)

4

Calm Before The Storm

As the conflict at sea developed, so the moment drew nearer when *Caroline* had to be ready to work with her sister light cruisers in the Grand Fleet. *Caroline*'s commissioning had yet to take place. Only when her builders were satisfied with her performance, after power trials, would the shipbuilder's ensign be replaced by the white ensign of the Royal Navy.

Both in German waters and British waters, the year 1914 saw stirring flashes of naval excitement, punctuating the interval between *Caroline*'s first keel plate laid still in peace time and her launch in September in time of war.

The creator of the Kaiser's navy, Grand Admiral Alfred von Tirpitz, had diligently made ready the German fleet for a conflict with the 'English'[1]. Tirpitz hoped an immediate clash would come about upon the outbreak of hostilities. Tirpitz believed that, although comparatively under-gunned, German built battleships had the edge over the English in quality of build, robustness, weight of armour protection and superior sub-division of compartments below decks. Knowing that the German High Seas Fleet was inferior only in numerical strength, Tirpitz's plan was to lure detached units of the English Fleet into German waters and destroy them with such fire power as was available from his heavy units waiting at their base in Wilhelmshaven in the Jade estuary. The strength of the English might also be sapped by deep-sea mining and by torpedoes launched from the newest of sea war vessels, the submarine.

Kaiser Wilhelm viewed strategy differently. It was his belief that the German fleet had to be kept safe, not exposed to risk. The fleet was the Kaiser's 'fleet in being', a tangible asset and useful as a bargaining counter should defeat loom. It was said that in Germany the Army was the senior service. Army generals, to Tirpitz's chagrin, decreed that it was best to keep the Navy intact. In the German high command the thinking was that it might not be necessary to interfere at sea with the transfer of a British Expeditionary Force to France. A Prussian backed juggernaut of an army would overwhelm England's puny land force.

1 How the enemy is referred to in *My Memoirs* by Grand Admiral Tirpitz

German naval units could now rapidly transfer from the Baltic Sea to the North Sea via the recently enlarged and deepened Kiel Canal in the event of a threat from the English fleet, or vice versa where the threat was from the Russians. The canal cut across the peninsula below Denmark between the German city of Kiel on the Baltic coast to the mouth of the River Elbe on the North Sea. It therefore would serve as a convenient transit route, saving fuel and time by avoiding the long route around Denmark to reach the North Sea via the Kattegat and Skagerrak straits.

Tirpitz believed that at the onset of hostilities the enemy would mount a close blockade of German ports and estuaries. This would frustrate German shipping movements. It would bring about shortages of food, and essential raw materials – a repeat of the English way of waging sea warfare against the French in the nineteenth century.

Paradoxically, in June 1914 there took place an event that was temporarily to banish thoughts of war by witnessing a moving example of the brotherhood of the sea, highlighting a healthy culture of competition between England and Germany. The occasion for this was the annual Kiel Regatta. This great sailing/yachting event had started in the usual way with the arrival of the Kaiser in his yacht, just as the most modern units of the German High Seas Fleet greeted a battle squadron from the British Grand Fleet with a courteous and warm welcome. Cocktail parties and dances were the order of the day. The English squadron had taken passage across the North Sea, made its way through the sound known as the Skagerrak, separating the southern tip of Norway from Denmark's mainland, whereupon it threaded its way south in between the Danish islands of Zealand and Funen, through the sound known as The Great Belt, to arrive in Kiel.

The mood was relaxed between the two rival seapowers. A German Korvetten Kapitän, Georg von Hase[2], equivalent to a commander in the Royal Navy, delighted in offering and sharing a clever wardroom toast with his British guests: "I drink to myself and an other. And may that one other drink to himself and an other – and may that one other be me."

However, at the sudden news of the murder on 28 June of Archduke Franz Ferdinand of Austria-Hungary, heir to the throne of the Habsburg Empire, Kaiser Wilhelm abruptly cut short his visit to Kiel. Just as the echoing report of a regatta starting gun will soon fade in the balmy, summer afternoon airs, so the fraternal mood of nautical fellowship quickly soured. What could the English squadron do? In the circumstances they made haste for home without delay, through The Great Belt to return to British home waters.

Home waters! Germany's advantage lay in the compactness of her fleet. She

2 Fire control officer of the *Derfflinger* by 1916

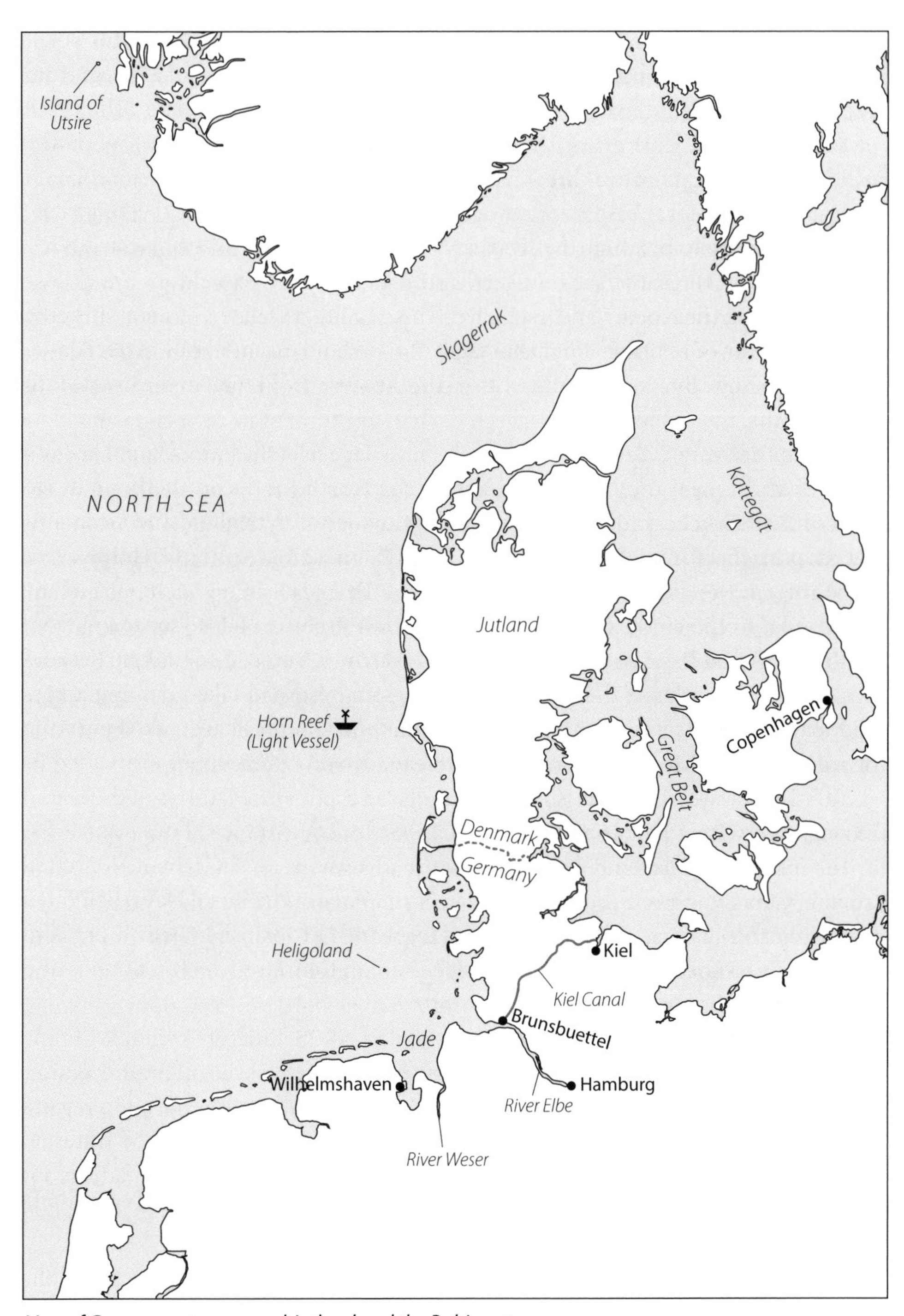

Map of German waters around Jutland and the Baltic entrance.

could consider the Baltic Sea fronting her northern and only coastline as her own province and reserve it for sea training. The Kiel Canal also helped to avoid the dispersal of her fleet.

By contrast the entire British Isles was bounded by seas on all sides. It was necessary in the light of the threat from the North Sea to rethink the usefulness of the traditional fleet war bases: Portsmouth, Portland, and Plymouth. And there were those bases in Ireland: Lough Swilly (Co Donegal), Berehaven and Spike Island (Co Cork), which covered Britain's western approaches. These bases were too distant from the arena, the North Sea, in which the next conflict would be fought. Britain's traditional war bases then might be used for harbouring her secondary fleet of older battleships. By blocking the exits to the Atlantic from the North Sea and the Dover Straits, the Admiralty had selected a battleground of its own choosing.

Where then might Britain's large modern surface fleet find more suitable bases from which to meet the threat from across the North Sea? Any challenge to the modern British fleet could only be promptly answered by having it conveniently placed somewhere suitable on the east coasts of England and Scotland. However, no harbour from Newcastle upon Tyne down to the Thames Estuary was large enough to play host to the entire Grand Fleet. The Admiralty was obliged to look farther north for a suitable safe anchorage. And so their need was addressed by a group of islands – the Orkneys. This bleak, windswept archipelago enclosed a circular, land-locked, expanse of sea, giving room enough for practice gunnery shoots and torpedo running. Moreover, it was accessible via narrow tidal passages.

This anchorage – Scapa Flow – at least offered a potential level of security that was not otherwise available on Britain's east coastline. A distance of some 580 miles lay between Scapa Flow and Germany's fleet bases. Work was underway by 1914 to protect Scapa Flow from surface assault and submarine attack. On Scotland's east coast two other bases were made ready: Invergordon in Cromarty Firth and Rosyth in the Firth of Forth. These bases were large enough to host one battle squadron with attendant battle cruisers. Lighter units such as light cruisers, destroyers and submarines were based at Harwich.

The time for courtesy visits, 'showing the flag', was over whether at home or abroad. Since the Balkan assassination of Archduke Franz Ferdinand, certain anxieties had sharpened the minds of Whitehall strategists.

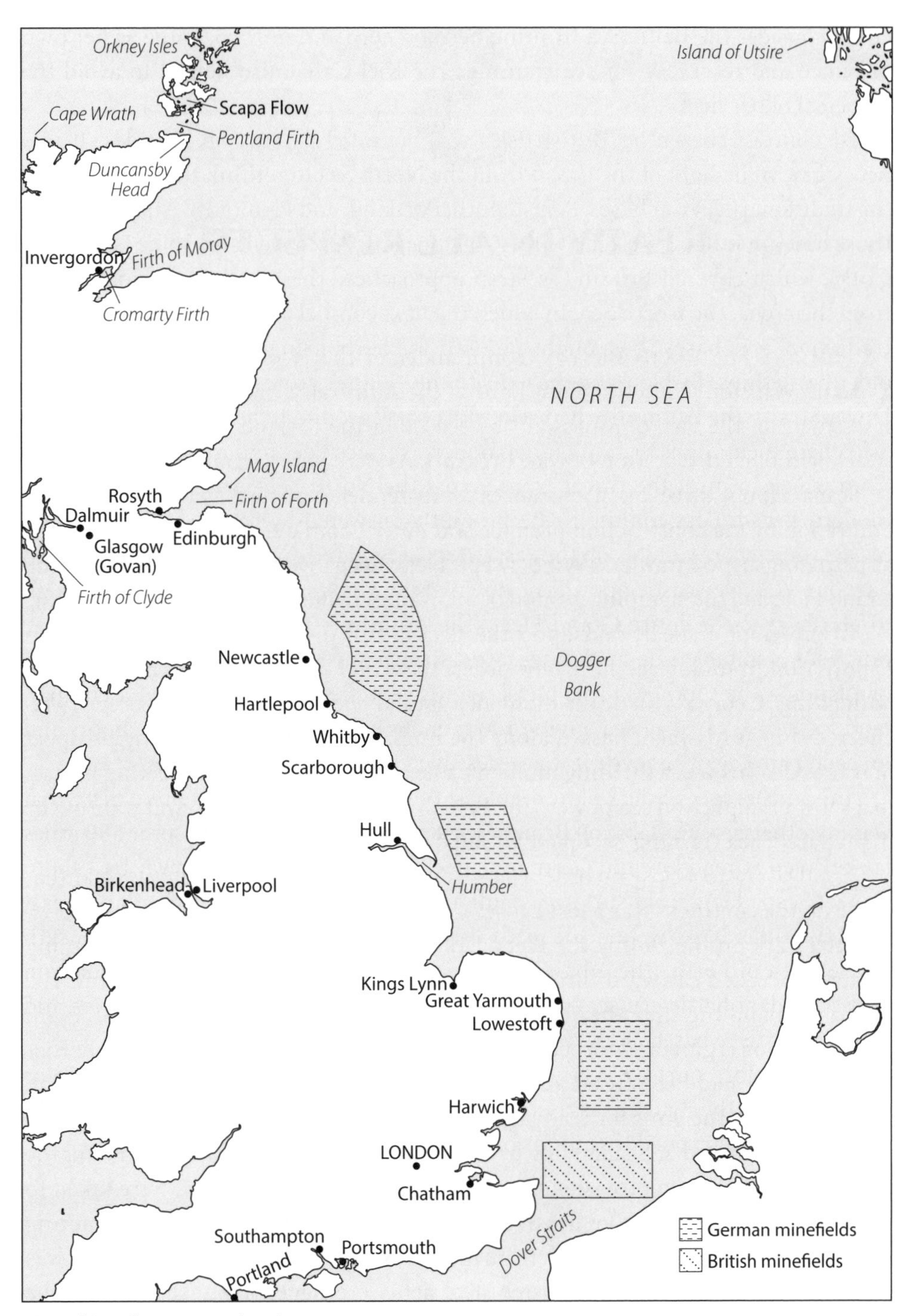

Map of British waters and main ports.

5

READY IN ALL RESPECTS

ADMIRAL SIR DAVID BEATTY, commander of the Navy's battle cruiser force and Winston Churchill, First Lord of the Admiralty, had together done a full assessment of the risk of an assault by sea upon their island nation – a surprise attack without a declaration of war. Precautions were put in hand.

Scapa Flow's three main entrances were rendered secure against attack and infiltration by the construction of inner and outer boom defences and the presence of patrolling armed trawlers. Wind swept, bleak Scapa would soon be home to the Grand Fleet and the stamping ground of many of its lighter forces including *Caroline*.

To pre-empt a sneak raid upon the fleet, a full mobilisation was ordered on 15 July 1914. It took place in review order, as in 1911, at Spithead. As last time the 'sailor king', George V made the moment a special one for the assembled crews, who cheered ship as his yacht passed along the lines. The grim, ever-hovering prospect of war aside, this was a thrilling moment, a flexing of muscle, hands beating breasts in a show of might and sea power. The British Navy would not be found wanting or ill prepared. Sea training occupied all ships for the next few days.

As July drew to a close no order was given to demobilise because it was clear that peace on the continent was a lost cause.

After dark on the 29 July the Fleet, led by some twenty dreadnoughts, keeping station proceeded eastward through the Dover Straits. No lights were shown. The next astern and the ship ahead appeared as black silhouettes. They were at 'night defence stations': guns and torpedoes loaded, the men sleeping by their guns. Earlier that day shells had been fused and warheads shipped on torpedoes. Ammunition was provided at the guns for ready use.

The Fleet arrived at Scapa Flow by dusk on the 31 of July. Coaling ship and the taking in of stores and provisions started at once. Further measures were taken to reduce the risk of injury from fire in the event of action and through the splintering of ships' boats and excess wooden furniture. All gear classed as 'fripperies' was culled, banished to the beach. Strong steel nets were rigged about the hulls of the big ships as protection from any surprise enemy submarine attack.

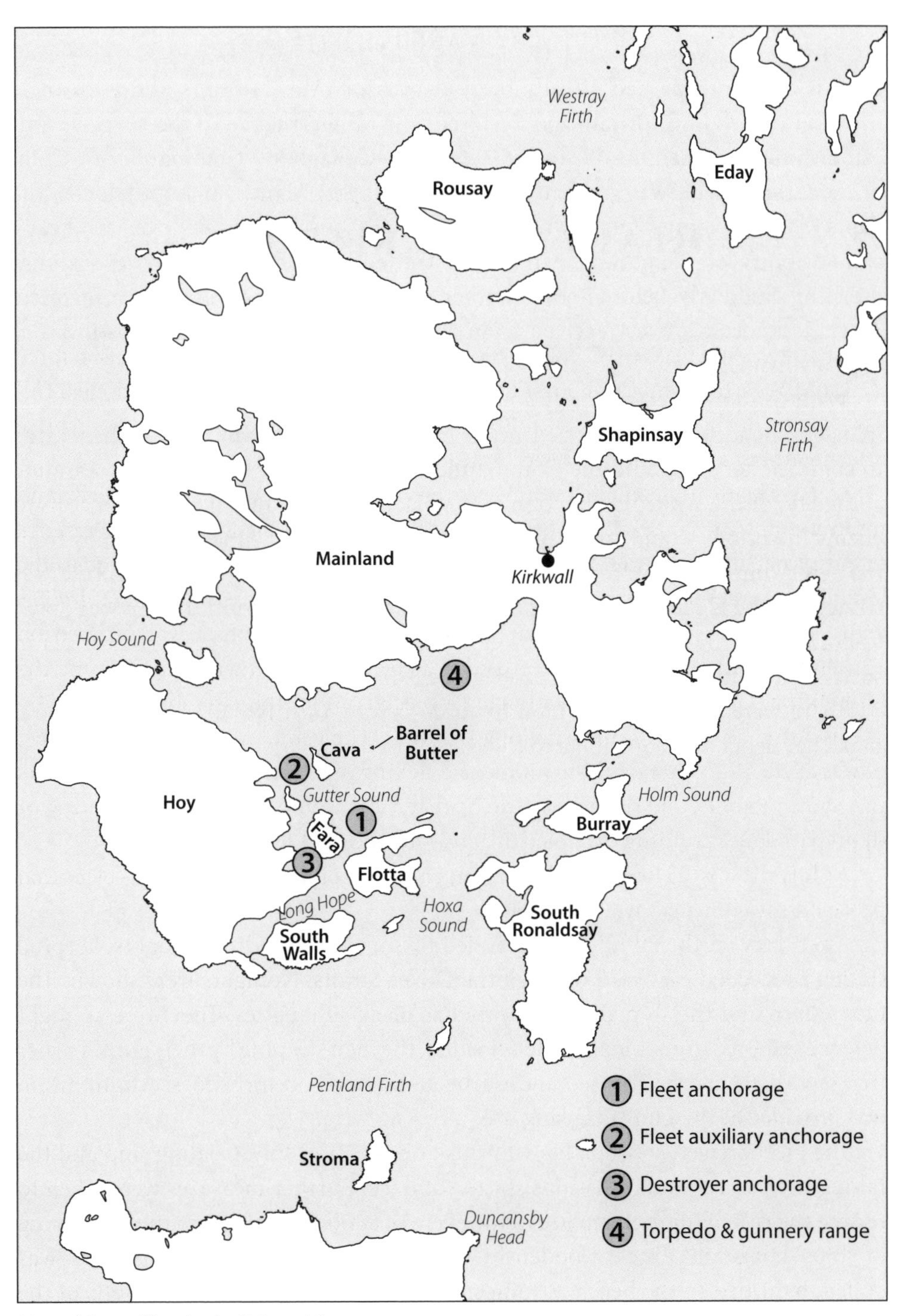

Map of Scapa Flow in the Orkney Isles.

TUESDAY 4 AUGUST, 1914

By this time Europe had already been at war for over 24 hours and with their invasion of Belgium, Britain too declared war on Germany. To the surprise and astonishment of senior officers, Sir George Callaghan, Commander-in-Chief of the Grand Fleet, was suddenly relieved of his command. Sir John Jellicoe, his deputy, and a younger man, was designated by Churchill to take over the heavy responsibility of Commander-in-Chief. The entire battle fleet put to sea that morning. Suddenly, before noon, another development. It was scarcely a surprise. A wireless message was received from the Admiralty: "Commence hostilities at once against Germany."

With this terse electronically conveyed message, the way sea warfare was to be managed henceforward changed in an instant. During many uncertain months to come, no longer would the Commander-in-Chief in his flagship carry absolute authority for the strategic direction of the Fleet. Instructions and intelligence as to enemy intentions would come to him from Whitehall, via wireless telegraphy from the new Admiralty building facing Horse Guards parade ground. If the future war strategy of the Fleet was ordained by the all-powerful Admiralty, then it was down to both John Jellicoe and David Beatty, who were afloat, to apply tactics at sea in such a way as to bring the German High Seas Fleet to action.

There were now just four months to go before Yard No 803 (*Caroline)* could make her debut and contribute to the war effort.

6

STEALTH AT WORK

A MUCH ANTICIPATED great battle between the German and British fleets did not take place on the outbreak of war. Instead, the open expanses of sea between the Faroe Islands, Shetlands and the Norwegian coast saw the start of a vigorous programme of patrolling by British forces to prevent exit of German shipping into the Atlantic, and its return home. The Dover Straits were similarly blockaded. These measures gave immediate effect to Britain's strategy of a distant blockade of Germany.

Nevertheless, between 6 August and 4 November 1914 the Royal Navy lost to enemy action seven cruisers and one very valuable newly built dreadnought, HMS *Audacious;* losses that gave meaning to Germany's strategy of attrition.

Kaiser Wilhelm's navy sought to avoid any situation that might lead to a decisive showdown, relying instead on achieving parity of numbers by making maximum use of an efficient sea mine and also the new invisible menace, her submarines. Some of these submersibles could range as far as the Pentland Firth, the North Channel and Britain's western approaches, positioning themselves by stealth off British havens, ready to use the element of surprise to exact deadly reprisals. For Germany it was a matter of deferring the challenge of battle until her navy was strong enough to win. It was a question of securing a 'strength equaliser', parity – *Kräfteausgleich.*

The British also were not slow to make use of submarines' invisibility. By clandestine observation in German territorial waters British submarines prepared the ground for the war's first serious surface conflict on 28 August 1914 – the Battle of Heligoland Bight. Amongst others, the action involved Admiral Beatty's battle cruisers and the light cruiser *Arethusa.* The media had their day hailing Admiral Beatty a hero. Defiant to the last, the German cruisers *Mainz, Ariadne*, *Köln,* and *V87,* a torpedo boat, were sunk.

Alexander Scrimgeour, a senior midshipman serving in HMS *Crescent* on the Northern Patrol, describes in his diary a typical foray on the 15 August: "Joined First Fleet at 7 am, 200 miles north of Wick. I don't quite realise what he [Admiral

de Chair] intended to use our 10th Cruiser Squadron for. The First Fleet look a very grim sight, and inspire the utmost confidence. We all proceeded over to the Norwegian coast, the battle cruisers leading, with the battleships four miles behind in three imposing lines, and light cruisers and destroyer flotillas scouring the seas all round on every quarter. There are no submarines … with us."[1]

This powerful force, arranged in 'cruising order', parallel columns, consisted of: The 1st Battle Squadron, eight dreadnoughts in the centre column, led by HMS *Marlborough*, Flag Ship, with HMS *Iron Duke* Flag Ship of the Commander-in-Chief in the van, 2nd Battle Squadron, eight dreadnoughts on the port flank, led by HMS *King George V*, 3rd Battle Squadron, eight dreadnoughts, led by HMS *King Edward VII*, on the starboard flank and as a back up the 4th Battle Squadron of four older dreadnoughts, including HMS *Dreadnought*. A mighty host, it consisted of twenty-nine capital ships. Right ahead, screening these dreadnoughts, steamed five parallel lines of destroyers, forty in total, each line led by a light cruiser. Ahead of the destroyer screen came the 10th Cruiser Squadron, and leading it were four light cruisers of the 1st Light Cruiser Squadron. Spearheading, twelve miles ahead of the main fleet, were the four battle cruisers of the 1st Battle Cruiser Squadron.

This routine 'sweep'[2] of the North Sea showed very early on in the conflict that Admiral Jellicoe meant to keep the sea, not merely with detached parts of the Fleet, but with the Fleet in its entirety. By so doing he could outnumber or at least equal any enemy force he might happen to encounter – a belt and braces approach.

It also shows the vital role being played by light cruisers as scouts in the overall dispositions of the Grand Fleet at sea. Although wireless telegraphy could aid communications at sea there was, of course, no electronic means, such as radar, by which an elusive enemy's whereabouts could be known in conditions of mist and fog. This was the nature of the environment in which *Caroline* would shortly find a place.

WEDNESDAY 4 NOVEMBER, 1914

A black day dawned at the Admiralty when dreadful news broke that Admiral Sir Christopher Cradock's battle squadron had been destroyed far away in the southern hemisphere, some fifty miles off the coast of Chile, at a location called Coronel. The two main stays of his task force, both obsolete cruisers: HMS *Good Hope* and HMS *Monmouth,* had been outgunned and outmanoeuvred by a faster, more modern squadron under Admiral von Spee. And to compound the sense of tragedy there

1 *The Complete Scrimgeour: From Dartmouth to Jutland 1913–16*, Alexander Scrimgeour (Bloomsbury, London, 2016)

2 'Sweep': the term was much in use to denote patrolling; no connection with minesweeping.

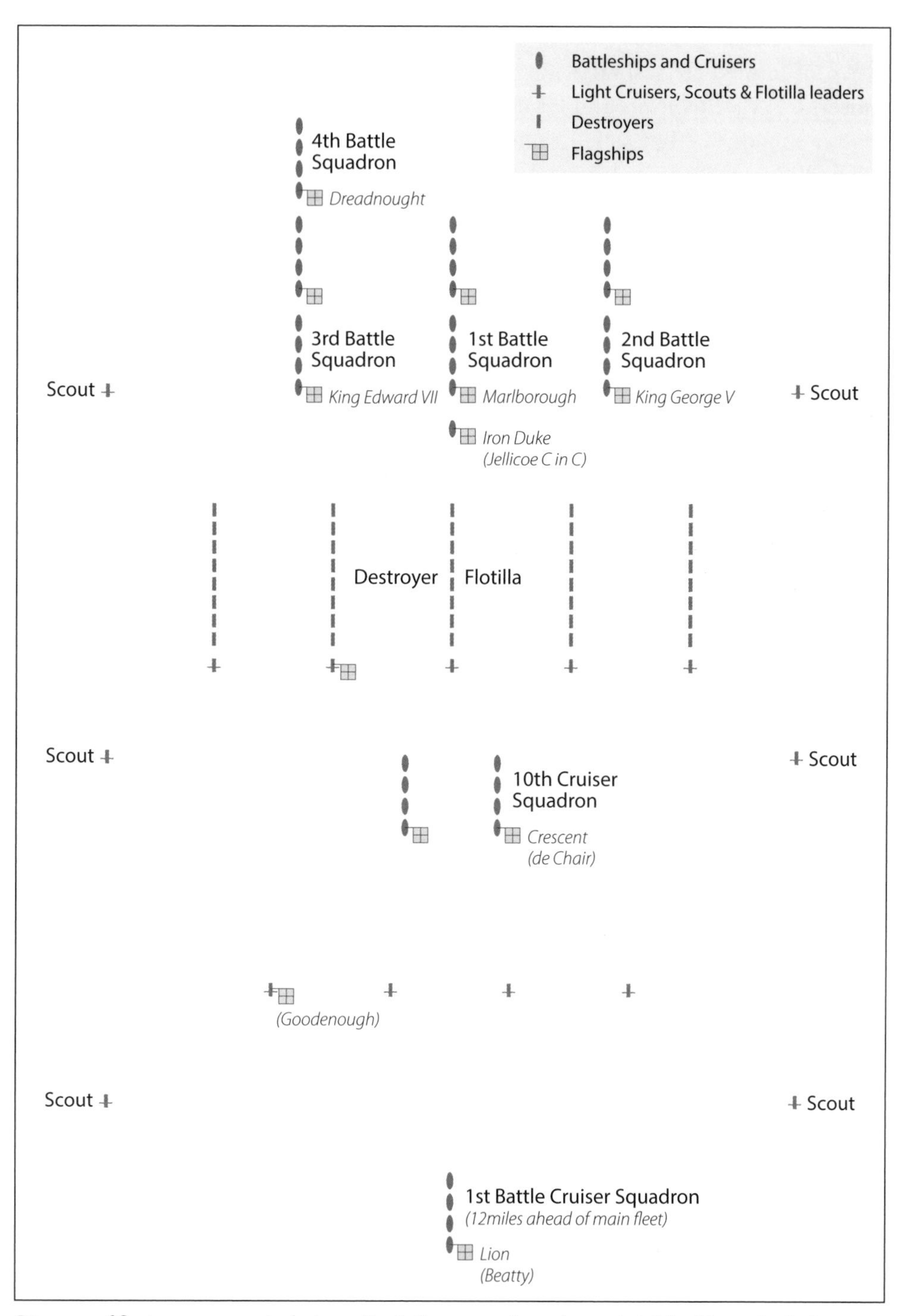

Diagram of fleet arrangements during a North Sea sweep, based on a sketch by Alexander Scrimgeour.

had not been a single survivor.

As an immediate response, First Sea Lord Admiral Fisher ordered two battle cruisers – HMS *Invincible* and HMS *Inflexible* – to be detached from the 2nd Battle Cruiser Squadron. They sailed from Devonport on the 11 November to avenge the defeat at Coronel. This recent and most serious blow to Great Britain's prestige as the foremost global naval power, badly dented her Battle-of-Trafalgar reputation for invincibility. Also a temporary absence in home waters of two valuable battle cruisers sent on a far distant mission to the bottom of the world, meant the German quest for *Kräfteausgleich* was almost a reality.

Admiral Beatty now had only four battle cruisers in home waters to Germany's five. Mines and torpedoes were taking their toll. The loss of *Audacious* plus the fact that two other dreadnoughts were refitting after a collision, severely reduced the Grand Fleet's margin of superiority over the High Seas Fleet. If it was ever the right time for the enemy to choose his moment to attack it was now. Indeed, Ernst Hashagen, one of Germany's submarine aces in this war at sea, held the view that British defensive material was of poor quality in the early days. British mines worked badly. They came to the surface at low water. So they could be seen from a long way off as little black dots, christened, 'caviar on toast' by the Germans.

Caroline had been launched on 21 September. In one month's time, on 4 December, her commissioning would take place...

7

CAROLINE – TWO PLUM APPOINTMENTS

FRIDAY 4 DECEMBER 1914

ALTHOUGH THE HUMILIATING defeat at Coronel was about to be reversed in equally dramatic fashion but this time to the Royal Navy's advantage, Lieutenant Commander the Hon ER Drummond had neither the time nor the inclination to dwell upon these developments right now. A regular Royal Navy commissioned officer, a two-and-a-half striper, Drummond had more immediate matters to attend to even as a final coat of grey paint was applied to his new charge lying in dry dock. From his time as a midshipman, sporting two white patches, to his next promotion to the rank of sub-lieutenant (one gold stripe with a curl) and subsequently to full lieutenant (two such with a curl), Drummond had progressed in the service well. He recalled a day back in October when he had received a large buff, 'official paid' envelope bearing the bold letters: OHMS – 'On His Majesty's Service'. In the course of his career he had received many of these. Some might contain an Inland Revenue demand, others might tell of a new appointment. The latter was much to be preferred. And so it proved on this occasion – a plum appointment to a brand new light cruiser, the equal if not superior to any current German light cruiser. It was worded in the Admiralty's inimitable, authoritative style:

> THE Lords Commissioners of the Admiralty hereby appoint you *Executive Officer RN* of His Majesty's Ship *Caroline* and direct you to repair on board that ship at *Birkenhead*… Your appointment is to take effect from the *1st November 1914*… You are to acknowledge the receipt of this Appointment FORTHWITH, addressing your letter to……… taking care to furnish your address.
>
> BY Command of their Lordships,
> Admiralty, S.W.1

It is usual for a senior officer in the regular navy to stand by to superintend the final stages in the completion of a warship – someone who has the experience to build a rapport with naval architects, dockyard foremen and technicians. This watching brief fell to Lt Cdr Drummond. Confirmation of his rise up the promotion ladder to the status of lieutenant commander had arrived in March 1914. It was a newly introduced rank denoted by two and a half gold stripes that bridged the gap between the ranks of commander (three gold stripes with a curl and the peak of the cap embossed with gilt oak leaves) and lieutenant. As the executive officer of *Caroline* he was her second-in-command. He thus had a heavy responsibility to his commanding officer for the fighting efficiency of the ship, its organization and routine. He was second then, only to the captain. Yet he had another function also, which was acting as the link between the *Caroline*'s other officers and her captain.

Earlier in the week when *Caroline*'s new captain arrived in the afternoon Drummond extended an invitation to him to meet the ship's officers in the wardroom, which was the mess exclusively reserved for the ship's commissioned officers. The commanding officer of a warship by contrast has the privilege of separate personal accommodation in deference to the weight of responsibility attaching to his command. However, as a guest of the wardroom the captain could get to know everyone in an informal atmosphere. What were the credentials of *Caroline*'s first captain? At 39, Captain Henry Ralph Crooke, RN, was a highly experienced gunnery professional. He was a product of Whale Island, home of naval firepower at Portsmouth, otherwise known as HMS *Excellent*. HR Crooke had experienced the ordered, disciplined atmosphere of Whale Island, where smartness in bearing and appearance soon became second nature to service personnel entering its precincts. Blue jackets passing through Whale Island learned quickly to obey snapped orders, to move 'at the double'. The quality of their parade ground drill was brought to perfection so that once they reached their ship they would turn out to be a credit to that ship and indeed the Senior Service[1].

Crooke's early naval service career coincided with key changes in attitude toward the science of gunnery. Great strides forward were being made at this time by increasing the rate of fire of naval guns and striving for improved precision and aiming accuracy over ranges from 5,000 to 10,000 yards. The man in the turret team, known as the 'gunlayer', became a critical figure in achieving best results. The personality behind this culture shift was Captain Percy Scott, who was selected to command the training school at Whale Island in 1903. Crooke certainly would have been inspired by this pioneer's insistence upon improving the rate of fire and its accuracy against towed targets when training at sea.

1 The nickname for the Royal Navy due to it being the oldest of the UK's armed services.

As a gunnery lieutenant 1st class in the *Calypso* from May 1896 to September 1897 Crooke put his personal stamp on this exacting and important discipline. A screw corvette, 2,270 tons, this ship's armament comprised four 6 inch and twelve 5 inch guns. A secondment to HMS *Excellent* in 1899 was followed by a sea going appointment to HMS *Repulse*, an eight-year-old battleship of 14,150 tons. Here Crooke could further build upon his reputation for expertise in the field of gunnery. His work earned praise from his commanding officers: 'all v.g., zealous and painstaking'. By 1905 Crooke had come to the attention of the Director of Naval Ordnance, John Rushworth Jellicoe, future commander-in-chief of the Fleet, who was then a captain. Captain Jellicoe described Crooke's work as, 'of the greatest possible value' and submitted his name for promotion. From March 1905 to April 1907 Crooke held an appointment as Assistant to the Director of Naval Ordnance (DNO). More praise followed: 'The DNO in reporting on the gunnery practices for 1906 brings to notice the services of Cdr Crooke to whom he attributes largely the work of re-sighting the guns and of fitting fire-control appliances, so satisfactorily carried out'. In February 1913 Crooke received expressions of appreciation of 'the zeal and ability displayed in carrying out trials of the Argo Range Clock in *Orion*'.

Now his career to date had reached a zenith with a plum appointment to a brand new light cruiser, HMS *Caroline*. Her main original armament included two superimposed single 6 inch guns aft in open turrets, one on the quarterdeck and the other on the deck housing. A total of eight 4 inch quick firing guns in the forecastle and waist, (four each side) plus twin 21 inch torpedo tube mountings placed amidships each side.

It was on this day as she lay in dry dock that the war log of HMS *Caroline* was commenced with the brief statement: *"At noon broke commissioning pennant..."*

8

Getting Underway – Take One

The commissioning of a warship is a masterpiece of organisation. Visually, the evidence of this is the breaking of a short, slender pendant at the ship's main masthead.

The men, referred to as 'ratings', who made up *Caroline*'s ship's company arrived at Birkenhead in the afternoon of Friday 4 December under the charge of Lieutenant WSE Gilchrist. Traditionally each warship has a home port, be it Plymouth, Chatham, or Portsmouth. *Caroline*'s association was with Portsmouth. Some men had been drafted straight from other ships; others might have wiled away the time at HMS *Victory*, Royal Naval Barracks, perhaps for weeks at a time, waiting 'for a draft', more might have been on leave or recently enlisted. So it was a matter of over two hundred ratings arriving at the dockside, each man with his bag, bearing personal hammock and clothing.

In joining their new ship many of these men were meeting one another for the first time. A good proportion of the crew might have hailed from Portsmouth or 'Pompey' as the city is affectionately known by sailors. But others came from more distant parts of the United Kingdom as, for example, James Weddick. At age 33, from Limerick in Ireland, and a Petty Officer Gunnery Instructor, his skill set was an invaluable asset to *Caroline*'s fighting efficiency. As each man stepped over the gangway, he was asked his name, which was ticked against a list. He was then given a 'station card'. It told him the number of his mess, his watch and part of ship, his action station, and his boat. Soon each man would find himself living and working with the same group of men, a sub group that would be divided into divisions which corresponded with their parts of ship. In the case of seamen these were: Fo'c's'le, Main Top and Quarterdeck[1].

Each division was divided into three watches: red, white, blue, arranged so that at any given time there might be a certain proportion of men on duty from each part of the ship. Thus it was that without delay the men were exercised at their duty

1 Terms derived from sailing ships and roughly equivalent to the areas in front of the bridge, amidships and at the rear of the vessel.

DIVISIONS

A daily routine, where the whole ship's company is split into the categories that correspond with their part of ship, (Fo'c's'le, Main Top, or Quarterdeck) and the men's speciality: 'supply',' electrical,' 'engine-room', 'signals'. The idea is to assemble the ship's company, 'muster' them, by inspecting them for fitness, alertness and smartness. It is an opportunity for the men to see their officers and the officers to see their men.

stations and also acquainted with fire and collision emergency procedures.

Then at the order 'clear lower deck' everyone mustered. They assembled in the stern of the ship on the quarterdeck to witness a traditional solemn rite peculiar to the fighting navy – the 'Commissioning' of their new warship. The occasion was the first time the men had an opportunity to meet the Captain and for him a chance to introduce himself by addressing the ship's company. It was the day on which Captain Henry Ralph Crooke RN formally accepted from her builders HMS *Caroline*, lying in dry dock at Birkenhead.

Finally the day ended as an armed guard was placed around the ship.

LOG BOOK

The foregoing only briefly touches on ship's organisation, so it may help to explain some of the terms used in *Caroline's* log. Selected abstracts of the log manuscript will appear in italics. Log entries are daily, whether in harbour or at sea, for the duration of the ship's commission. The text is concise, terse and matter-of-fact being a record of the ship's position, movements, prevailing weather and how the men and the ship were employed. A log may also record details of loss or damage to stores, and disciplinary action. From these abstracts an attempt is made to entwine HMS *Caroline*'s own role in the war at sea with that of its general historical context.

Caroline's log books up to the battle of Jutland are in three strongly bound HM Stationery folios. The first covers 4 December 1914 to 28 February 1915. The second runs from March 1915 to April 1916. The third begins on 1 May 1916 and ends in 1917. The log books were first opened to the public in 1966.

Saturday 5 December 1914

AM *Overcast, cloudy, squally, wind: West North West, 6–7[2], glasses not shipped*
09.00 divisions, stations for going out of harbour
09.15 commenced flooding dock, hands preparing ship for sea

PM *00.45 took sentries off their posts, proceeded out of dock under tow from tugs, ship with a list to port of 10°*
1.30 came to port anchor, 6 shackles, in Mersey off ship yard
2.15 weighed, commenced to swing ship for adjusting compasses, however, due to contractors being unable to raise sufficient steam to stem the tide this had to be abandoned
4.15 came to starboard anchor 6 shackles, 13 fathoms off the yard
6.00 exercised general quarters
7.45 secured, exercised darken ship

Once assisted into the frothy, turbid waters of the Mersey while a strong wind blew from the west, *Caroline*'s clinometer recorded a list to port of 10°. We note also that, as yet, no supply of mercurial barometers (glasses) had been delivered.

The issue of correcting the list was put on hold for the present, the main objective being to 'swing ship'. This, however, did not happen, perhaps because too few of her eight Yarrow boilers were 'flashed up' to raise a sufficient head of steam to confront the Mersey's swift tidal stream. The ship might still have had some Cammell Laird engineers on board.

SWING SHIP

For navigation a standard Admiralty liquid compass was used, placed on the fore-bridge with a good all round view. On board a steel vessel the magnetic compass will not point to Magnetic North. It will be affected by the magnetism of the iron in the ship. The difference between the direction of Magnetic North and the direction of North as shown by the compass on board is known as the 'deviation of the compass.' The amount of error will depend on the direction of the ship's head. The ship was swung by transit marks with Vauxhall Chimney in this instance. The difference is recorded for different directions of the ship's head, the results being tabulated on a board. The stand containing the compass or 'binnacle' was made of wood with brass fittings. Certain magnets and soft iron fittings were used as correctors in adjusting the compass.

2 Reference to wind speed, i.e. 6–7 (strong breeze to high wind) on the Beaufort Scale.

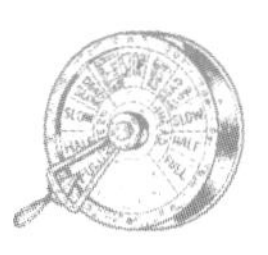

DARKEN SHIP

In time of war darken ship takes place at dusk – all scuttles, portholes and hatches are closed so that no artificial light can be seen from outboard. Scuttles are also made watertight and portholes made black out proof by applying hinged 'deadlights' – a shutter fastened over the window, also used in stormy weather.

9

GETTING UNDERWAY - TAKE TWO

Sunday 6 December 1914 – From Mersey to Clyde at sea

AM *07.30 weighed, swung ship for compasses in the Mersey*
08.45 proceeded to Bar Light Vessel, 10 knots, manned A1 and B1 guns
09.15 250 revs
10.00 general quarters, Bar Light Vessel abeam to port ½ miles
10.25 shaped course N 50° W[1]*, 300 revs*
10.45 carried out gun trials
11.15 streamed log, a/c[2] *North, 11.20 a/c West, 11.45 N 50° West*
12.20 ceased fire

PM *12.30 worked up to full speed, 25,000 HP exercised night defence stations*
3.00 Calf of Man S 73° E, 2½ miles
4.00 eased gradually to 250 revs, darkened ship
8.00 Ailsa Craig abeam 3½ miles

THE DAY ON which the light cruiser made her maiden passage began with a gentle to moderate westerly wind, 3 to 4. The sky was mainly blue, at times cloudy. Later the wind veered, easing to a gentle breeze.

The earliest possible moment was taken to try out and exercise the ship's armament. By 10.00 they came to 'general quarters', a state of operational readiness in which a part of the watch is at their action station. The ship's speed through the water was obtained by streaming a patent log from the stern. On her way to the Clyde Ailsa Craig, the volcanic outcrop off the Ayrshire coast, was observed three and a half miles on the starboard beam.

1 Ships course, i.e. North 50° West (North being 0°, East and West 90° there from and South 180°)
2 i.e. altered course

STATES OF READINESS

Fourth degree: General Quarters/Cruising Stations – a part of the watch is closed up at their quarters.

Third degree: Defence Stations – one watch or part of a watch, according to the proportion of the armament required to be manned, closed up.

Second degree: Action Stations relaxed – all hands closed up except a small proportion from each quarter may 'fall out' (dismiss) for refreshment.

First degree: All hands closed up as action with enemy imminent.

PATENT LOG

This is an instrument that is trailed or streamed from a ship's rail at the back of the ship, i.e., from its stern. Made of brass, it contained a rotor, the blades of which revolved a set number of times every mile. A dial on the instrument recorded the revolutions in miles – the distance run through the water. The ship's speed through the water was then noted from the elapsed time from the start of the operation to its termination.

Monday 7 December 1914 – From Mersey to Clyde at Dalmuir

AM *07.00 lit fires in remaining boilers*
08.30 worked up to full speed, carried out trial runs on measured mile
10.50 proceeded 350 revs up the Clyde, exercised general quarters
11.40 came to starboard anchor 10 fms off Greenock, stations for going in and out of harbour

PM *3.00 weighed, 4.00 under charge of pilot, lost overboard torpedo davit port side aft by fouling corner of basin*
6.25 made fast alongside Derbent *in Beardmore Basin, Dalmuir. Let all fires die out except A1 boiler*

Tuesday 8 December 1914 – In basin at Dalmuir, From Dalmuir to Govan

AM *09.15 divisions, guns' crews at divisional drill*

PM *1.00 lit fires in 3 boilers*

1.30 struck down all ammunition from upper deck, received 701 lbs fresh meat, 800 lbs vegetables
3.45 proceeded up river to Govan
5.05 made fast alongside North Wall Princes Dock, let fires die out in all boilers except B1, special leave to watch till 07.00

Wednesday 9 December 1914 – At Govan in Princes Dock

AM *09.00 divisions*
10.00 guns' crews at divisional drill

PM *1.30 leave to boys till 19.00*
2.35 cast off, proceeded into dry dock
3.45 made fast in No 2 graving dock
4.00 hands employed in shoring up ship
6.00 ship docked, shored up, special leave to watch till 07.00

BOYS

Of *Caroline's* total ship's company of 289, 13 were boys. Their duties in harbour were to keep the gangways clean, take messages and man the side as required (side boys). At sea they acted as bridge messengers. Others were known as Call Boys who were stationed about the ship to repeat all pipes (instructions conveyed by bosun's whistle) so that all parts of the ship heard the instruction. This method of passing orders was later replaced by the introduction of broadcasting loudspeakers. No 47 of Wills' Cigarette Cards series, 'Naval Dress and Badges', gives the following description of the function of these, nowadays considered under-age, members of the crew: 'Boys are entered into the Navy between the ages of 15¾ and 16¾ or as Youths from 16¾ to 18. They must be strong, sound and well developed and agree to serve until they are 30. They are trained first on shore or in harbour ships, then passed into the Home Fleet, completing their training in the Fourth Cruiser Squadron. Pay starts at 6d a day and they may rise to Lieutenant.'

Thursday 10 December 1914 – At Govan No 2 Graving Dock

AM *09.00 divisions*
10.00 guns' crews at divisional drill, search light crews divisional drill
12.00 winter clothing issued

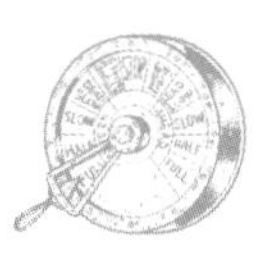

PM *3.15 stations for night defence*
5.45 darkened ship
6.10 exercised night defence stations till 19.45
9.00 exercised night action stations
9.30 secure

Friday 11 December 1914 – At Govan No. 2 Graving Dock

AM *08.00 watch muster for medical inspection*
09.00 divisions
09.25 exercised general quarters for 1 hour afterwards exercised control parties and sightsetters

PM *1.00 discharged to hospital 2 ratings*
2.00 guns crews at divisional drill

Saturday 12 December 1914 – At Govan

AM *02.00 commenced to flood dock*
06.15 ship's company landed on jetty to enable stability test to be carried out
08.00 ship's company returned on board
08.50 undocked, shifted berth
10.00 lit fires in three boilers
10.30 made fast alongside Plantation Quay
10.40 HMS Owl *came alongside starboard side, pumped 65 tons of oil from our tanks into those of* Owl

PM *4.15 HMS* Owl *shoved off*
5.50 cast off, proceeded to Princes Dock
7.10 made fast No 6 berth
8.00 let fires die out in all except one boiler

Sunday 13 December 1914 – From Govan to Greenock

AM *09.00 all hands except working party out of the ship*
10.00 divisions, prayers
10.30 carried out stability trials
11.00 lit fires in 3 boilers

PM *1.25 finished stability trials, hands employed in getting out pig iron ballast on to jetty*
2.30 cast off from jetty
3.00 proceeded down river under charge of pilot, 180 revs, exercised general quarters
4.10 300 revs
4.30 secured, darkened ship
4.50 night defence stations 180 revs
5.10 stopped: came to starboard anchor off Tail of the Bank, 7 shackles, 17 fms
8.00 anchor bearings: Rosneath Beacon, Whitefarland and Princes Pier

---

Monday 14 December 1914

Caroline's *draught before trial – fore 13 ft 0 in, aft 14 ft 11 in*
Easterly wind, overcast, cloudy
Displacement on leaving anchorage Firth of Clyde, 3955 tons, barometer 28.7" Weirs, mean draught 13 ft 11½ in

AM *05.00 lit fires in all boilers*
07.05 weighed
07.45 proceeded down harbour 250 revs
08.00 gradually work up to 400 revs
08.55 commenced working up to full speed, course and speed as requisite for running full power trials on measured mile, commenced trial runs on measured mile at 25,000 HP
10.00 exercised general quarters
10.40 commenced 4 hour trial at 30,000 HP and above

PM

2.40 finished full power trial, draught after trial: fore 12 ft 7 in, aft 14 ft 4 ½ in, 165 tons fuel oil remaining

3.00 proceeded to anchorage off Greenock

4.00 alongside oil tanker Derbent *off Greenock. When coming alongside starboard cathead securing plate was bent and 12 ft length of* Derbent's *three part rail and stanchions was bent or broken plus her top strake on its upper edge*

5.45 2 prisoners apprehended and brought aboard, 1 placed in irons below, blacksmith, bosun's mate, 2 shipwrights repair damage

8.30 finished oiling

8.45 cast off proceeded to anchorage

9.30 came to port anchor 11 fms 7 shackles

Tuesday 15 December 1914 – At Firth of Clyde

AM *06.00 at single anchor Greenock, calm wind ESE*

07.30 lit fires in all boilers

08.00 unrigged night defence gear

09.00 divisions, physical drill, marines at infantry drill, guns crews at divisional drill. Hands re-stowing provision room and preparing for sea

10.45 shortened in

11.00 weighed proceeded down to the Firth

PM *0.30 worked up to full power*

1.05 commenced 4 runs on measured mile at full power

1.10 finished trials, eased down gradually

1.20 carried out turning trials

1.50 swung ship, compass adjusted off Greenock

2.50 proceeded to anchorage

3.00 came to port anchor, 11 fms 7 shackles

3.45 oil tanker Derbent *alongside*

4.30 night defence stations, darkened ship, let fires die out except one boiler

7.15 burned searchlights and exercised crews

Wednesday 16 December 1914 – At Greenock

AM *04.00 finished oiling, 796 tons*
06.00 at single anchor, light airs
09.00 divisions, physical drill
10.00 guns crews at divisional drill

PM *3.00 took in 3 tons coal*[3]
4.00 prepare for sea, night defence stations
7.00 burned search lights

---

3 The galley range used coal for catering.

10

North About To Scapa

THURSDAY 17 DECEMBER 1914

CAROLINE'S STABILITY HAD been corrected. Two days of full power trials proved satisfactory, echoing her successful launching at the year's start. Her armament tested, *Caroline* had ammunitioned. They had completed with fuel, stores and provisions. The 'duffel coat' made its debut as the crew received winter clothing.

They were now bound for the bleak Orkney Islands and Scapa Flow, where daylight came at about ten in the morning and darkness at about half-past three. Very often in these northern isles the sky might be heavily overcast with low driving clouds. Mostly in Scapa it blew hard. *Caroline*'s heavy weather baptism awaited her as she shaped a course for Cape Wrath via the Western Isles of Scotland and the sea passage known as the Minches. Cape Wrath, the headland at the extreme westernmost tip of northern Scotland, was her waypoint eastward to the Pentland Firth, a fast flowing narrow stretch of sea separating northern Scotland from the Orkney Isles.

A smooth sea on the measured mile was all well and good but the new light cruiser would now roll, pitch, plunge and wallow in the leaden-coloured, heaving seas in these fastnesses. On her next plunge sheets of spray and spindrift would sweep over her fo'c's'le drenching the officer of the watch and lookouts on the open bridge. Nor had her raw ship's company the time to settle down, find their sea legs. Envious newcomers to life at sea in the Navy had noticed how the old hands seemed immune from sea sickness, how their faded shore-going pale blue jean uniform collars contrasted so markedly with the new recruits' deep blue newly issued flannel collars. In time after much scrubbing, rinsing, dhobeying, maybe a light azure blue shade could be attained to endow its wearer with the mark of an experienced sailor. At Scapa a period of 'working up', would interrupt periods of boredom. Daily exercises in seamanship and gunnery would soon bring the ship's company to a peak of fighting efficiency and ready to deal with every possible emergency.

Meanwhile in Scapa on Tuesday 15 December tedium brought on by an

apparently inactive enemy was suddenly broken by a signal from the Admiralty. Four battle cruisers, five light cruisers and about thirty destroyers had put to sea from the River Jade. With two of Admiral Beatty's battle cruisers away in the South Atlantic hunting down von Spee the German high command had seen an opportunity for an 'equaliser', the long sought *Kräfteausgleich*; something to offset the morale-lowering effect of their defeat 8,000 miles away. At this stage the Admiralty did not know the enemy's plan.

Always it had to be borne in mind that the Imperial German Navy could spring a surprise attack with their full might whenever they chose to do so. The thought continually nagged at Admiral Jellicoe that his foe, having come out, might try to draw him over a minefield and submarines. Just now it was essential to have at his disposal a numerical superiority over the enemy.

But just as a football team's strength may be weakened by injuries so Jellicoe's battle fleet had to contend with maintenance problems arising from their being ever required to keep the sea: the need to not only clean boilers but also to re-tube them, deal with faulty condensers and turbine problems. It took weeks to repair ships after unfortunate collisions at sea.

Be that as it may, the emergency squadron at Scapa, the 2nd Battle Squadron of eight dreadnoughts, already had steam up. So it put to sea at once while Beatty's battle cruisers, light cruisers and destroyers sallied out from Cromarty and Rosyth with the aim of not only bringing the enemy to action but also to place strong forces between the enemy and his base. It was in the words of the First Lord of the Admiralty, Winston Churchill, a "tremendous prize… the German battle cruiser squadron whose loss would fatally mutilate the whole German Navy and could never be repaired – actually within our claws…"

On Wednesday morning 16 December at 08.00 the enemy's motive became clear – a tip and run raid on England's east coast. That old acquaintance from the 1911 Spithead Review, SMS *Von der Tann,* this time visited the holiday resort of Scarborough, attacking the town by bombarding it from the sea. *Von der Tann,*

KEEP THE SEA

To maintain a successful maritime offensive, e.g. mount a blockade of an opponent's harbours as in the Napoleonic wars and the Great War, it was necessary to organise round the clock patrols of the seas. In the Great War these patrols often involved the entire Grand Fleet and became known as 'sweeps'. It was expensive both in terms of fuel and wear and tear of ships. But it kept the sea safe for Great Britain and discouraged the German Navy from risking the ships of High Seas Fleet.

joined by SMS *Derfflinger*, began a devastating assault of the town with their secondary armament at a range of just one mile. The first salvoes straddled the massive 300 foot high castle headland protecting the town. Extensive damage was done to the heart of Scarborough, shells striking the Olympia Picture Palace and the Grand Hotel. Eighteen citizens were killed, and hundreds more wounded. Within some thirty minutes the undefended town had been subjected to a taste of indiscriminate warfare. In making their approach the raiding force had made use of a gap in a minefield that they had earlier laid off the Tyne and Humber.

Also bombarded were Hartlepool and Whitby. Leading the detached raiding force was the German light cruiser SMS *Kohlberg*. This light cruiser's mission was to head south and lay one hundred mines off Filey.

Jellicoe had realised the enemy would make his escape by that gap in the minefield. He added the 3rd Battle Squadron to the hunt. The prospect of cutting off the German force from their base looked good – until 11.00. That was when the weather suddenly changed. It became misty and partly as a result of poor weather and a lack of lucid signalling by the British forces the wheel of fortune turned to neither side's advantage. Next day the British Press echoed the public's indignation: 'Where was the British Navy?' In a letter to the Mayor of Scarborough, however, the First Lord of the Admiralty had the last word: "Whatever feats of arms the German navy may hereafter perform, the stigma of the baby killers of Scarborough will brand its officers and men while sailors sail the seas."

While HMS *Caroline* was independently making her way north to Scapa the events in the North Sea would have been made known to Captain Crooke via wireless telegraphy. Now, the cry went out on many a recruiting poster: 'Remember Scarborough!' The effect was to harden the resolve of those in the Navy to get back at the enemy. *Caroline*'s ship's company would have been in a state of heightened awareness as to the danger from enemy mines and roaming German submarines.

It fell to upper-deck crew members to face and to endure the angry elements and the raking shell fire of the enemy in action. Equally dangerous was the lot of boiler room staff: stokers and engine room mechanics toiling below the waterline. With only the thickness of three inches of armour plate as a shield between them and the inrush of the sea, a mine or a torpedo might seriously alter the design of their machinery-filled habitat into a terrifying shambles of twisted metal and scalding steam. Efforts to control damage in compartments and to keep a ship from foundering could be hampered by dead and injured shipmates and the escape of high pressure scalding steam from fractured pipes. Up top on deck men had a chance to save their lives – for those at their posts deep below the waterline their chances were fair if in a lucky ship blessed with a zealous and painstaking captain.

The log takes us now to Scapa:

Thursday 17 December 1914/ Friday 18 December 1914 – Greenock to Scapa Flow

AM *03.40 sighted Cape Wrath Lt. [light] N 70° E*
08.30 course as req. for entering Hoxa Sound
10.07 came to port anchor, 12 fms, 6 shackles in Gutter Sound

PM *1.45 reported enemy sub in Scapa Flow*
2.05 proceeded at highest speed through Hoxa Sound to eastern entrance of Holm Sound
2.40 500 revs, zigzagged up towards Copinsay without sighting the enemy so returned to Scapa via Hoxa Sound.

It was arguably a jittery response by the ship's captain to the submarine sighting report. New to his command perhaps it was the right time to play 'captain caution', as we shall soon see. Other than having recourse to high speed ramming and gunfire, there was no effective defence against an enemy submarine at large at this early stage in the war[1]. This particular alarm, however, proved the seriousness with which Captain Crooke appreciated the threat and the sharp alertness of *Caroline's* look-outs. There were, of course, false alarms a plenty. A seemingly black fairway buoy in the twilight, with the tide swirling past it like a wake, could be mistaken for an enemy periscope. Even the presence upon the surface of the sea of an innocent seal could lead to an excusable error by a look-out.

1 Soon after the declaration of war a German submarine had penetrated the Firth of Forth. All day she observed the movements of British warships through her periscope. At night the boat surfaced to replenish its air supply. For the remainder of the night it lay on the bottom, giving the crew sleep and rest.

11

With The Grand Fleet

SATURDAY 19 DECEMBER 1914

At dawn just after their arrival, the ship's company witnessed an inspiring sight. It was the business-like entry into the Flow of the First Battle Squadron: *Marlborough, Collingwood, Colossus, Hercules, Neptune, St Vincent, Superb* and *Vanguard.*

This was all part of the fleet's regular sea training routine. Like Nelson, Admiral Jellicoe believed in the value of constantly training the Fleet for war. Repeated practice in gunlaying meant that this vital function might be performed in record time. The complete squadron had been to sea to perform heavy firing exercises, full calibre shoots in the western entrance of the Pentland Firth. The drill called for a practice target to be towed by a ship, the *King Orry.* Acting as an invigilator was the destroyer *Oak.* She was a tender to the Commander-in-Chief's Flag Ship, *Iron Duke* and carried senior officers whose duty it was to mark and record the results.

That afternoon *Caroline* found herself not in the open sea but in a position some two and a half miles eastward of a tiny skerry within the Flow called the Barrel of Butter, for gunnery practice on a sub-calibre range. On weekdays cruisers and destroyers followed each other for training, performing gun and torpedo practices. The Flow was a hive of bustling warships 'working up'; preparing for war.

In the process of working up we see from *Caroline*'s log that the procedure was not as straightforward as might have been desired. The log records a sortie by the Grand Fleet over the Christmas period. Determined not to be unpleasantly surprised, as at Scarborough, the Commander-in-Chief decided to keep the fleet at sea to meet and destroy any of the enemy venturing out at this festive time.

Sunday 20 December 1914 – At Scapa Flow

AM *07.45 weighed, when heaving in port cable parted, anchor and swivel piece and 1 fm of chain lost overboard*

07.55 enemy sub attack signalled, recd. orders from C in C to anchor again
08.35 Came to starboard anchor, 15 fms

Monday 21 December 1914 – At Scapa Flow

AM *08.30 weighed, shifted berth ½ cable, dropped buoy over estimated posn of lost anchor, came to starboard anchor 14 fms, 5½ shackles. A drifter and two whalers away sweeping for lost anchor and cutter with divers looking for same.*

PM *2.00 searching for anchor*
4.00 1 Reindeer hair buoy and 2 sinkers ½ cwt to mark anchor on departure
10.40 weighed and proceeded out of Hoxa Sound at 240 revs

Tuesday 22 December 1914 – Scapa to Cromarty then up to Invergordon

Read warrant No.2

Wednesday 23 December 1914 – Invergordon

Read warrant No.3, discharged to Duke of Edinburgh *prisoner for detention*

READING WARRANTS

'Clear Lower Deck' is the Royal Navy's order for all members of a ship's company to muster (i.e. to assemble) in a convenient space to hear a special announcement, for example an urgent change to the ship's programme or about the possibility of going to 'action stations'. In *Caroline*'s day this order was also used where a punishment warrant was read aloud to the entire ship's company in the presence of the accused. This was a document or an instrument adding authority to the details of a charge under the Naval Discipline Act for which a rating had been found guilty. Such a charge would be of a very serious nature, issued by the captain and approved by a flag officer. By the act of reading the charge to the assembled ship's company everyone on board knew what crime had been committed and what punishment the offender would receive.

Friday 25 December 1914 – At sea, wind SSW 3–4

AM *06.45 sighted battle fleet, a/c S 28° E*
07.00 sighted destroyer flotilla to starboard
07.50 worked up to full speed
08.00 hands at action stations
10.00 a/c N 74° E
10.45 co. S 45° E, zigzagged 2 points each side of course for 10 minutes on each course

PM *4.00 took station astern of* Galatea, *2 cables astern* Galatea

---

Saturday 26 December 1914 – At sea, wind SSW 7

AM *02.15 lost sight of* Galatea
03.15 sighted Falmouth
08.00 commence to zigzag 2 points each side of course, 10 miles on each
11.30 wind 8 to 9 south by west
11.55 man overboard, course and speed as req for picking up

PM *0.45 man not recovered, Edward James Morris, Pte RML, lost overboard Kisbie lifebuoy 1 in number*
4.00 took station astern Galatea, *force 9*
8.00 wind SSW 4–5

ROYAL MARINE LIGHT INFANTRY

'The physical standard for the Marines is higher than for any other regiment outside the Life Guards. The headquarters of the 'Red' Marines, as the RMLI are called, are at Chatham, Portsmouth and Plymouth and numbers are carried in all the larger ships of war. Kipling's name of "soldier an' sailor too" applies to them very well, for parties are frequently landed for shore service.'

Taken from Wills's Cigarettes, 'Naval Dress and Badges' Series.

MAN OVERBOARD

In the most severe weather conditions prevailing lowering a boat to save the man was not a practical course of action for his recovery. 'Course requisite for recovery' could be interpreted as the ship turning round and retracing the course steered in the hope of sighting him. A 'Kisbie' lifebuoy was thrown over to assist the man in the water. The main danger to anyone who falls overboard comes from being struck by the propeller – a danger naturally greater in a quadruple-screw ship.

On 27 December *Caroline* returned, not to Scapa but to Cromarty and the naval base at Invergordon, having covered 264 miles. Two days later a Court of Inquiry was held on the death from drowning of Private James Morris, aged 30 from Walsall. This man, a member of the complement of Royal Marine Light Infantry in *Caroline,* was the ship's butcher. Quarters of fresh beef to be conveyed to the galley were most likely stored behind a screen on the open deck, access to which was dangerously risky in vile weather.

Meanwhile, further north at Scapa, as the 2nd Battle Squadron approached the base to enter during a gale and on a dark night, the *Conqueror* collided with the *Monarch* in the narrow entrance to the Flow. Both ships sustained heavy damage and were put out of action, pending repairs. With the *Audacious* already sunk, the squadron was now reduced to only five units.

THURSDAY 31 DECEMBER 1914

Back in August 1914 the British public had felt that by all that was reasonable the war would be over by Christmas. Instead by New Year's Eve the tempo of the war at sea was gradually intensifying, though no huge clash between the opposing battle fleets had as yet occurred. Tragically, as we have seen within this short time span the cruisers: *Amphion, Pathfinder, Hawke, Hermes, Aboukir, Cressy* and *Hogue* had all made the ultimate sacrifice.

Captain Crooke to date, during the short tenure of his first command at sea in a brand new light cruiser had taken no chances, had exercised sound judgement as an officer of the newer school; a man determined not to be caught unawares by the unseen enemy's submersible beast of stealth and prey, the U-boat. Even as the traditional Royal Navy custom was enacted at midnight when the youngest of *Caroline*'s ship's company rang sixteen bells in lieu of the usual eight bells to ring the Old Year out and the New Year in, a tragedy was unfolding in the English Channel.

It was suggestive of a navy operating at a two-tier level of security in response to

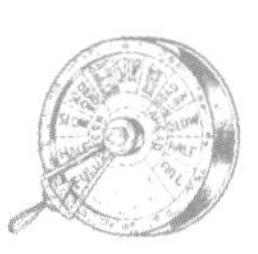

the submarine menace. In the north a healthy, grim respect for the 'unterseeboot' reigned in the Grand Fleet, born of learning lessons the hard way. In the south in the Channel Fleet the idea had taken root that the channel sea area was beyond the range of the U-boat and so the threat could be discounted in those waters. It was proof, alas, that the unwary older school officer even now still existed – in complacent denial of the menace of the U-boat. And so it was that on the last day of the year a Vice Admiral, who was Commander-in-Chief of the Home Fleet, operating in the Channel, believed there was no danger from U-boats, as there had been no reports of any operating in his area, the western part of the English Channel. In the early hours of the 1 January the 5th Battle Squadron was on a training exercise near Start Point, off Portland. Without zigzagging the squadron was following a straight course at only 10 knots under a bright moon. The last ship in the line, HMS *Formidable* was stricken by two submarine-launched torpedoes with the loss of thirty-five officers and five hundred and twelve men.

It was almost as if the squadron had asked for trouble by presenting itself as an easy target. Indeed even the offer of a destroyer escort had been declined by this senior flag officer on the ground that: "I had not the slightest idea that the Channel was 'infested' with submarines."

TIME

The day was split into seven parts, called watches, using the 24.00 hours' system.

Midnight to 0400 Middle watch, 0400 to 0800 Morning watch, 0800 to 1200 Forenoon watch, 1200 to 1600 Afternoon watch, 1600 to 1800 First dog watch, 1800 to 2000 Last dog watch, 2000 to 0000 (Midnight) First watch.

The time during a watch, except during the quiet hours, was marked by striking a bell every half-hour, a tradition going back to the use of a half-hour glass.

Thus by the end of a watch eight bells was struck, as for example at Midnight, 0400, 0800 and so on – hence the striking of 16 bells to herald in the New Year.

12

CAROLINE ESCORTS HOME A SAVAGED *LION*

THE OPENING DAYS of the New Year 1915 found *Caroline* persevering with gunnery practice, using a towed target, it being the turn of the gunners manning the 4-inch battery on the ship's port side. The hands were paid their quarterly settlement. After the 'man overboard' incident, earmarked for special attention were the ship's Kisbie life buoys, which were tested and found correct. Hammocks were, as described in the log: *scrubbed and washed.* Then:

Sunday 3 January 1915 - At single anchor, Invergordon

PM *2.00 joined ship Capt. C.J. Wintour, 1 captain's steward, 1 valet, 1 cook*
4.00 left ship Capt. Crooke

The log omits to add detail about the members of the crew, apart from the captain, who 'left ship' this day. It would seem that an exchange of personnel between HMS *Swift* and HMS *Caroline* took place. CJ Wintour held the appointment of Captain of the Fourth Destroyer Flotilla, serving in HMS *Swift* at Cromarty. *Swift* was a larger than usual destroyer, a flotilla leader, 2,175 tons. As such the destroyer flotilla leader had extra accommodation for a destroyer flotilla Captain and his staff. It is likely that Captain Wintour's temporary appointment to *Caroline* was for the purpose of completing the training exercises that went into working up the ship to maximum efficiency, to the level of the other light cruisers in the squadron. We shall come across Captain Wintour later as the story of Jutland unfolds with all of its tragic consequences.

Meanwhile, in the 'lower deck', amongst the ratings who joined the *Caroline* on commissioning day, 4 December 1914, was a leading steward, Albion Percy Smith. Born in Tunbridge Wells, Sussex on 16 January 1890, Smith had a successful merchant navy career behind him, as an officers' steward. He had been with the

THE HAMMOCK

On leaving the training ship every boy was given two hammocks, i.e. rectangular white canvas mats, one to use and one as a clean replacement after two weeks. The boy marked his hammocks with his name, using 1-inch block, black painted letters. To 'sling' his hammock and to 'lash' it up he was also given one complete set of strings. This consisted of 32 lengths of three-strand white hemp cordage, known as 'clews' plus two short lengths of cordage known as 'lanyards'. He was given one other length of cordage known as a 'lashing'. The clews and the lanyards enabled him to suspend his hammock from hooks in the ship's deck head. 16 clews were attached to the 'head' of the hammock and 16 clews to its other end known as the 'foot'. A lanyard at the head and at the foot was secured to the hooks. When the hammock was not in use he would 'lash' it up neatly and 'trice' it to the deck head. In scrubbing a hammock the crew member used soap and a hand scrubber and re-scrubbed with salt water. Clean hammocks were slung once a fortnight. The bag and the hammock were kept until he left the Service. Hammocks and bedding were aired once a week by placing and securing them over the ship's guard rails.

Image from A Seaman's Pocketbook, *HMSO, 1952*

Bucknall Line, serving in the SS *Johannisberg* on the Southampton/South Africa passenger route up to October 1914. The month of November saw him taking induction training at HMS *Victory*, Royal Naval Barracks, Portsmouth. Men serving in the Royal Navy were rated according to skills they had and tasks they performed, hence the term 'rating'. Captain Crooke now chose Smith to accompany him as his personal steward whilst seconded to *Swift* for this month. It was to be the start of a good, useful working association between captain and manservant, one that was to ripen in lasting friendship and the exchange of Christmas cards in the years to come. To return to the ship's log:

Monday 4 January 1915 – At single anchor Invergordon

AM *10.00 Exercised collision stations and abandon ship stations*

Tuesday 5 January 1915 – At single anchor

Recd on board 1 in no. stockless 5 cwt anchor in lieu of anchor lost at Scapa Flow, Dec. 20th

---

Wednesday 6 January 1915 – Cromarty Firth, calm

AM *10.00 Carried out .303 practice from 6 inch guns at towed target over Nigg sands*

PM *1.00 Exercised ammunition supply parties and control parties*

Thursday 7 January 1915 – Cromarty Firth

AM *10.00 carried out .303 practice from stbd guns at towed targets*

Friday 8 January 1915

PM *6.30 carried out night defence practice with blank .303 at towed target*

Between Saturday 9 January and Wednesday 13 January 1915 the ship's company were exercised ashore on route marches each day. On 14 January ship's boat crews were exercised 'away under oars', the remainder of the hands 'painting ship'. On 16 January (with Captain Wintour still on board) 6 inch supply parties were exercised at drill and gunlayers and sightsetters under instruction. On 18 January the men were exercised at 4 inch ammunition supply parties and 6 inch guns' crews at loader.

Warrant No 4 was read on 19 January. Also on this day a zeppelin raid took place on Great Yarmouth and Kings Lynn.

Thursday 21 January 1915 – Invergordon and at Sea

AM *10.25 speed as reqd. for carrying out firing (full charges) with 2nd and 3rd divisions*

PM *12.10 ceased fire*
1.20 increased to 20 kts, speed and course for firing torpedo at target

towed by Achates, Ambuscade *having picked up* Caroline*'s torpedo came alongside and retnd. same.*

During *Caroline*'s working-up period at Cromarty the German Admiral Von Ingenohl explored the idea of another sortie. The Kaiser approved the plan in so far as it could be executed without hazarding the main German battle fleet of dreadnoughts, which were his 'fleet in being.' It was not that the German Admiralty had set themselves a New Year resolution to mount an offensive, but that they felt sufficiently bold after the Scarborough show to make a reconnaissance in force toward a relatively shallow area in the North Sea known as the Dogger Bank, some 180 miles west of Heligoland and roughly 60 miles east of Southwold. The intention might have been to interfere with British fishing trawlers and thus to surprise, trap and destroy isolated elements of the Grand Fleet, perhaps account for a British battle cruiser or two, units of which they believed frequented the area. Stationed further south now at Cromarty rather than at Scapa and therefore nearer the scene of a possible conflict, would *Caroline* see a chance of action?

So it was that at 17.45 on 23 January Admiral Franz von Hipper's force put to sea: three battle cruisers, one armoured cruiser, four light cruisers and eighteen destroyers. With just three battle cruisers: *Seydlitz*, (25,000 tons) *Moltke* (23,000 tons) and *Derfflinger* (26,600 tons), Hipper could well have been aware that he was under-strength in comparison to his enemy Admiral Beatty, who could deploy straight out of Rosyth up to five modern battle cruisers. Hipper's armoured cruiser, *Blücher* (15,850 tons), though a product of the finest, most robust Teutonic ship construction, lacked speed and would, in a chase at sea, fall behind her more modern sister ships.

Conspicuously missing this time was Britain's old acquaintance of Spithead and Scarborough memory, *Von der Tann*, detained in dockyard under repair. The Grand Fleet's enemy, so competent, so disciplined at the science of gunnery, skill in ship handling and seamanship had yet to learn a new discipline – wireless silence. Decoders in Whitehall, recognizing the call signs of Admiral Hipper's ships could work out the strength of his task force by eavesdropping on enemy wireless broadcasts. Hipper's exact objective, however, was not clear.

In some excitement the First Sea Lord, Sir Arthur Wilson, strode into the Old Admiralty House board room, a favourite haunt of Winston Churchill's in Whitehall, exclaiming: "So those fellows are coming out again."

The battle that occurred a little north of the Dogger Bank turned into a chase of the German ships and saw the British force ineptly concentrate its fire power upon the comparatively slow *Blücher*. This gallant ship went down, all guns furiously

defiant to the end. At the same time an exchange of fire had resulted in severe damage to the *Seydlitz,* and Beatty's flag ship *Lion* had received hits below her waterline and damage to her generators. So badly damaged was *Lion* that Beatty had to leave her and transfer his flag to another ship in order to continue directing the chase. But an outright victory by the British had been missed due to the misguided obsession of focusing gunfire exclusively upon the already mortally damaged *Blücher.* It enabled the three remaining modern German battle cruisers to escape.

Shortly after the action the German Navy launched an inquiry into the damage to *Seydlitz.* She had narrowly escaped instant destruction. A projectile charge had caught fire inside a damaged turret. But for prompt safety action by a brave crew member, the flash from it might have exploded ammunition in the magazine store located far below. To make projectile charges less flammable the Germans henceforward encased them in brass. Tragically this design flaw – the need to stop a flash of flame bolting like lightning downward from a turret to the magazine – was not corrected on the British side. Instead silk cordite bags continued in use. We can see from the log how the aftermath of this action played out:

Friday 23 January 1915 – From Invergordon to Sea

AM *08.50 slipped and procd down harbour with 4th flotilla, 10kts*
09.20 passed boom defence, 12 kts
09.35 a/c South 70° East, 18 kts

Saturday 24 January – At sea

AM *07.45 went to action stations & cleared away*
08.00 sighted battle fleet 1 pt. abaft port beam
09.00 a/c South 75° East to close
09.45 took station 10 miles ahead of Flag at 19 kts
10.00 In station
11.00 a/c to avoid floating mine and again at noon. TBD's[1] *from flotilla told off to sink these*

PM *5.45 a/c south having sighted* Lion *towed by* Indomitable
6.00 In station on stbd beam of Indomitable
11.00 [HMS] Broke *and 2nd flotilla joined up*

1 A class of vessel called Torpedo Boat Destroyers or, often, just Destroyers.

Sunday 25 January 1915

AM *07.10 exd station keeping astern of* Lion, *3 miles*

Monday 26 January 1915 – From Sea back to Cromarty Firth

AM *03.00 parted company with* Indomitable *&* Lion

By 27 January *Caroline* had returned to Invergordon. Scarborough avenged, Dogger Bank was hailed as a success by the British Press. The enemy had quit the field of battle at the loss of the *Blücher* and damage to the *Seydlitz*. Such a verdict was, however, not enough to still Admiral Beatty's restless hunter's instinct. Outright victory had eluded him. A signal that said: "attack the rear of the enemy" had been taken to mean "finish off the *Blücher*", and that to Beatty's intense disappointment, had led to the chase of the remaining three top prizes being inexplicably broken off. The hunter had sought nothing less than four scalps. What he had collected was just one scalp. Worse still a lesson in 'health and safety' was effectively ignored.

13

WAR – 'A PERIOD OF INTENSE BOREDOM PUNCTUATED BY MOMENTS OF INTENSE FEAR'...

ON SATURDAY 30 January 1915 Captain Wintour left the ship to return to *Swift*. Captain Crooke and four domestic ratings, including Albion Smith, returned to *Caroline*. Any notion that the captain's personal steward enjoyed a more privileged status than others of the ship's company was long dispelled by now. Captain Crooke was a most punctilious commanding officer, a creature of routine insisting on the highest standards of tidiness and cleanliness in his quarters. It fell to Smithy, as he was now known to his shipmates, to keep order in this domain to ensure that the captain's seven pipes, seven razors and seven toothbrushes, one for each day of the week, were arranged in as orderly a fashion as a Guard of Honour. The quarters comprised a night cabin with adjacent bathroom and lavatory, a day room/cabin and a sea cabin abaft the ship's bridge and next to the chart house. Should the clock on the bulkhead of the captain's sea cabin not show the right time, Smith was 'put in the rattle' for carelessness. So differential treatment for the captain's steward was out of the question. As Smith writes: "although I was his personal man he was stricter with me than anyone else on board – he was really a marvellous man and I would have gone anywhere with him."

A look at the ship's log for the first days of February 1915 shows that in the absence of enemy sorties the crew were kept busy and being trained to the necessary state of operational readiness.

Sunday 31 January 1915 – From Invergordon to Rosyth

PM *4.30 secured for sea, darkened ship*
5.40 slipped from buoy, prod down harbour at 250 revs
7.00 a/c S 65° E, zigzagged every 5 mins, 25° either side, 340 revs...

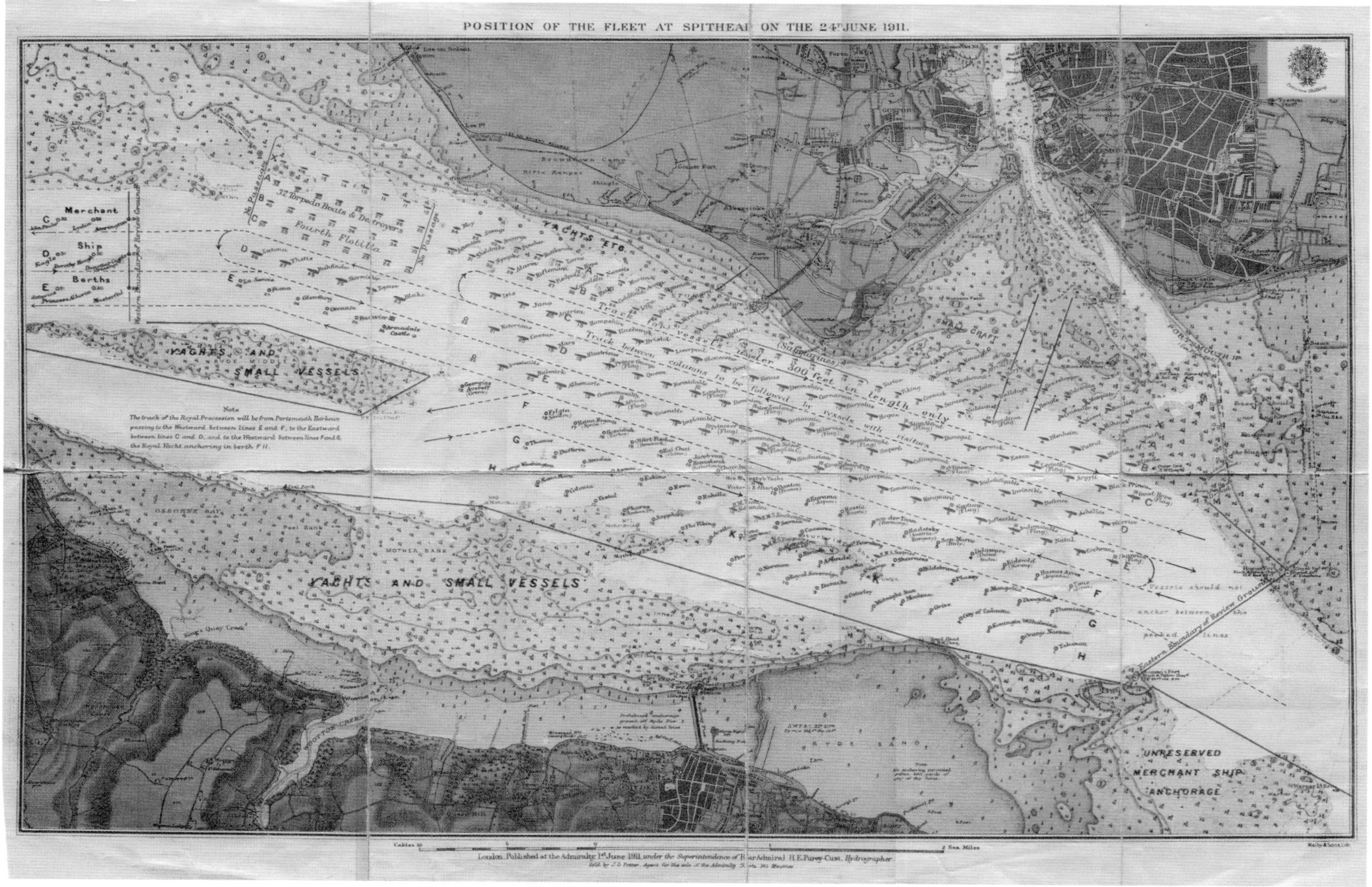

POSITION OF THE FLEET AT SPITHEAD ON THE 24TH JUNE 1911.

Review of the fleet at Spithead, 1911. An original colour chart by Gale & Polden.

Author's Collection

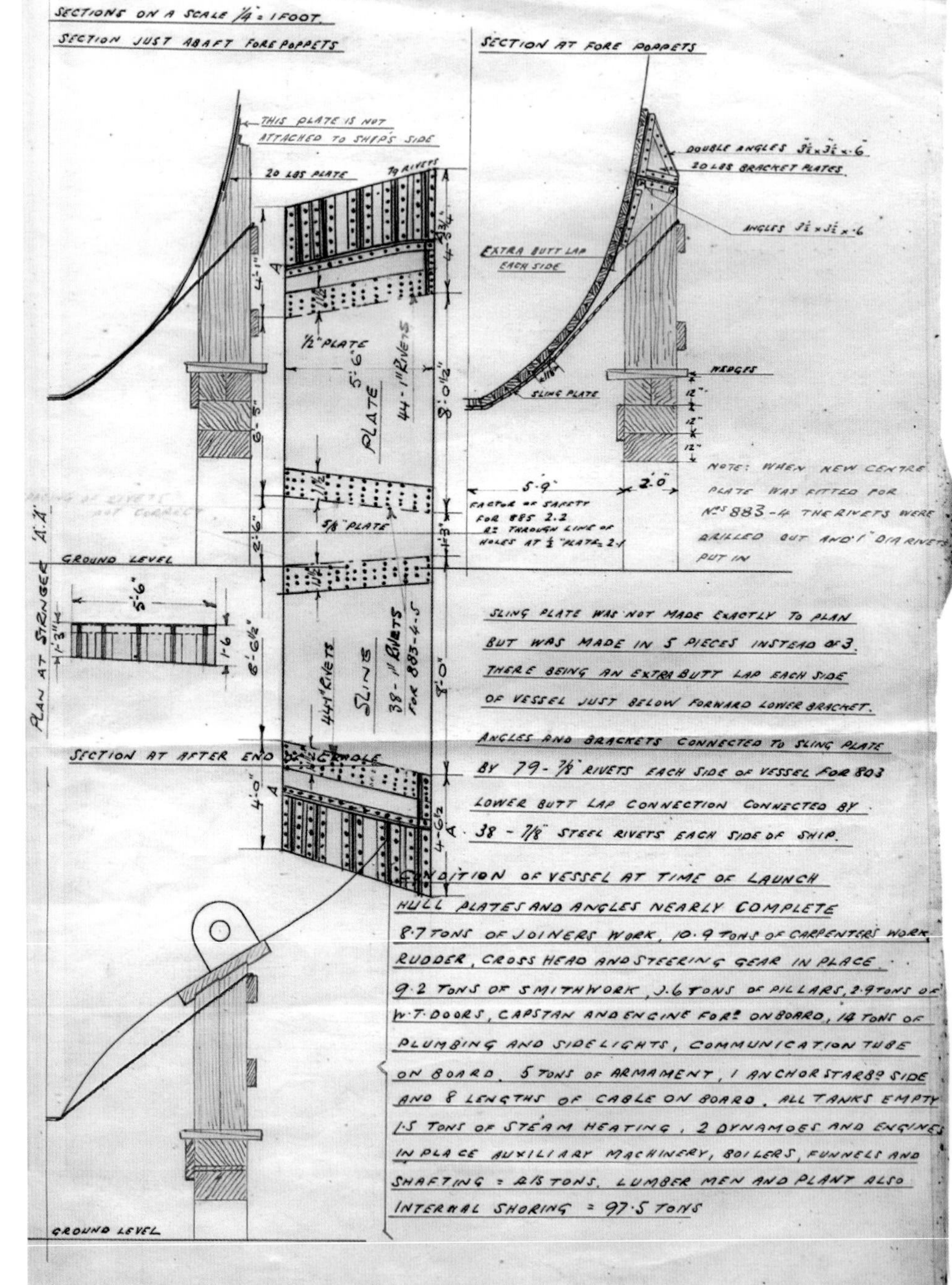

H.M.S. "CAROLINE"
LAUNCHING ARRANGEMENT
SECTIONS ON A SCALE 1/4" = 1 FOOT.
SECTION JUST ABAFT FORE POPPETS
SECTION AT FORE POPPETS
THIS PLATE IS NOT ATTACHED TO SHIP'S SIDE
20 LBS PLATE
DOUBLE ANGLES 3½ x 3½ x ·6
20 LBS BRACKET PLATES
ANGLES 3½ x 3½ x ·6
EXTRA BUTT LAP EACH SIDE
½" PLATE
PLATE
44 - 1" RIVETS
WEDGES
SLING PLATE
NOTE: WHEN NEW CENTRE PLATE WAS FITTED FOR NOS 883-4 THE RIVETS WERE DRILLED OUT AND 1" DIA RIVETS PUT IN
5/8" PLATE
GROUND LEVEL
PLAN AT STRINGER 'A.A'
SLING
44 - 1" RIVETS
38 - 1" RIVETS FOR 883-4-5
SLING PLATE WAS NOT MADE EXACTLY TO PLAN BUT WAS MADE IN 5 PIECES INSTEAD OF 3. THERE BEING AN EXTRA BUTT LAP EACH SIDE OF VESSEL JUST BELOW FORWARD LOWER BRACKET.
ANGLES AND BRACKETS CONNECTED TO SLING PLATE BY 79 - 7/8" RIVETS EACH SIDE OF VESSEL FOR 803
LOWER BUTT LAP CONNECTION CONNECTED BY 38 - 7/8" STEEL RIVETS EACH SIDE OF SHIP.
SECTION AT AFTER END OF CRADLE
CONDITION OF VESSEL AT TIME OF LAUNCH
HULL PLATES AND ANGLES NEARLY COMPLETE
8·7 TONS OF JOINERS WORK, 10·9 TONS OF CARPENTERS WORK,
RUDDER, CROSS HEAD AND STEERING GEAR IN PLACE.
9·2 TONS OF SMITHWORK, 3·6 TONS OF PILLARS, 2·9 TONS OF
W·T·DOORS, CAPSTAN AND ENGINE FORD ON BOARD, 14 TONS OF
PLUMBING AND SIDELIGHTS, COMMUNICATION TUBE
ON BOARD. 5 TONS OF ARMAMENT, 1 ANCHOR STARBD SIDE
AND 8 LENGTHS OF CABLE ON BOARD. ALL TANKS EMPTY
1·5 TONS OF STEAM HEATING, 2 DYNAMOES AND ENGINES
IN PLACE AUXILIARY MACHINERY, BOILERS, FUNNELS AND
SHAFTING = 815 TONS. LUMBER MEN AND PLANT ALSO
INTERNAL SHORING = 97·5 TONS
GROUND LEVEL

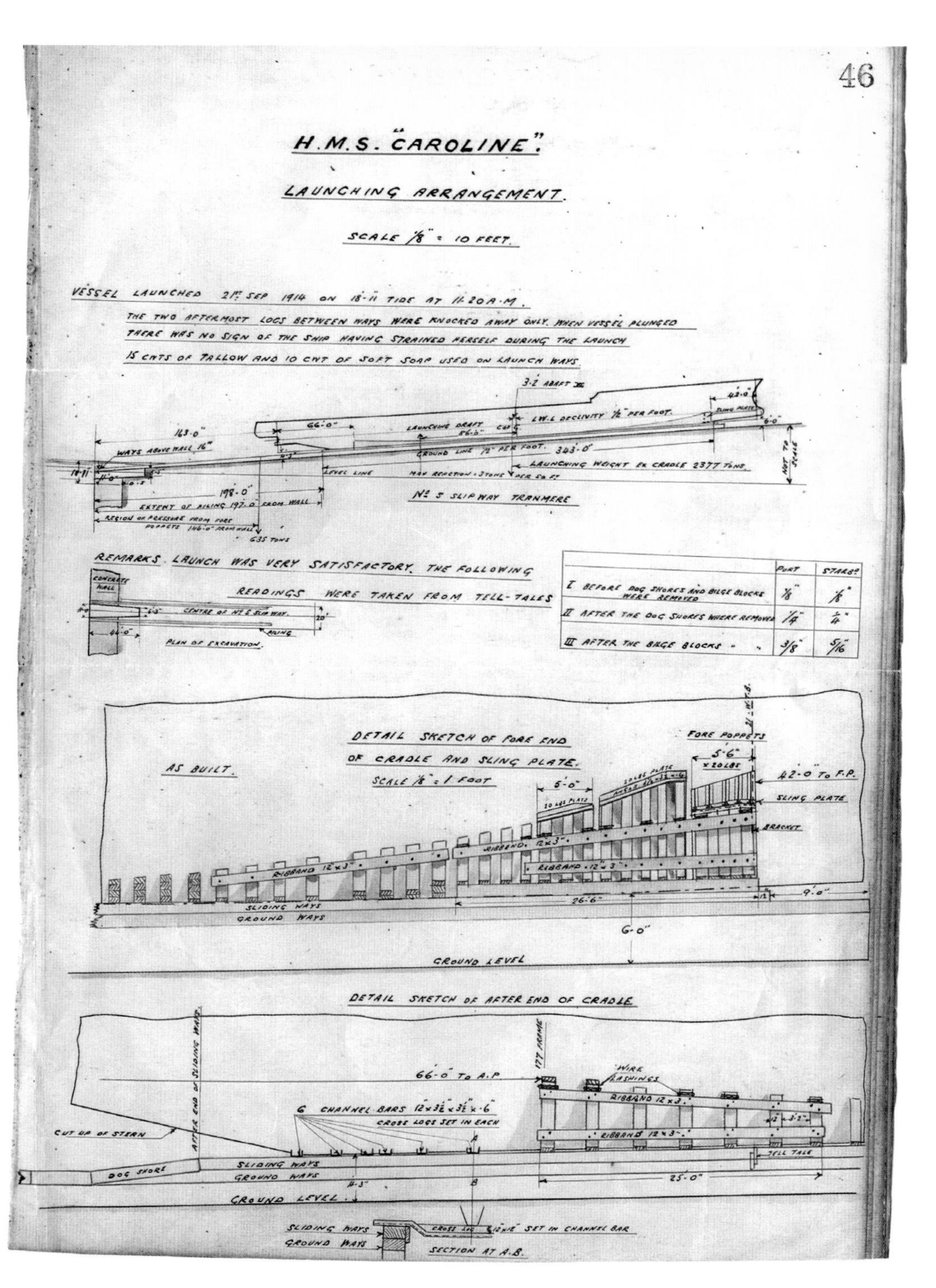

Launching arrangements for Yard No 803, HMS *Caroline*, at Cammell Laird Shipbuilders, Birkenhead.

Courtesy Wirral Archives Service

LIGHT CRUISER, 1913-1914 ESTIMATES.

BUILDERS. CAMMEL-LAIRD No 803 JAN.'14 - DEC '14

DIMENSIONS. 446' (oa) 420' (bp) × 41½' × 13-16' NOMINAL DISPLACEMENT 3750 TONS, (3800 FULL LOAD).

ARMAMENT. F'CASTLE. 4×4" Q/F GUNS REPLACED 1917-18 BY CENTRELINE 6" BL & 2×3" H.A.
WAIST. 4×4" Q.F GUNS REPLACED 1917-18 BY CENTRELINE 6" B.L.
TWIN 21" TORPEDO TUBE MOUNTING AMIDSHIPS EACH SIDE.
3 pdr H/A GUN AFTER ENGINE ROOM CASING (REMOVED 1918-17) 2×3 pdr SALUTING GUNS EACH SIDE
AFT. 2×6" BL GUNS (SUPERFIRING)

PROTECTION. 3" SIDE AMIDSHIPS, 2¼"-1½" BOWS, 2½"-2" STERN, 1" UPPER DECK AMIDSHIPS & RUDDER HEAD

MACHINERY GEARED TURBINES, 8 YARROW BOILERS, 4 SCREWS. 30,000 HP = 28½ KNOTS 917 TONS FUEL OIL
NAME SHIP OF CLASS

SERVICE RECORD

GRAND FLEET (4TH FLOTILLA) 1915.
1ST & 4TH LIGHT CRUISER SQUADRONS 1915-18
EAST INDIES FLEET 1919-21
PAID OFF – DEVONPORT – 1922

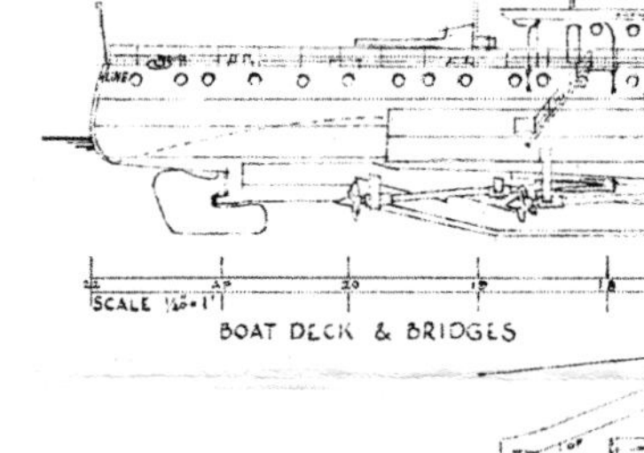

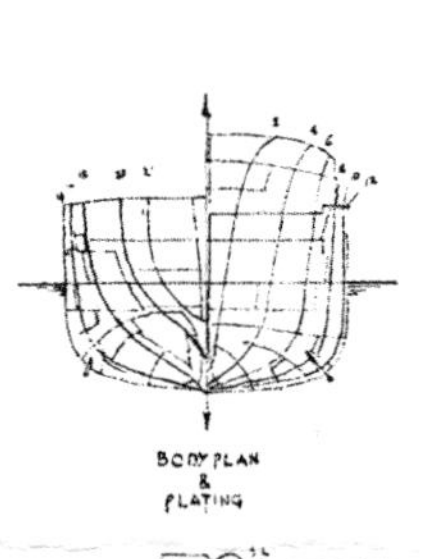

SCALE

BOAT DECK & BRIDGES

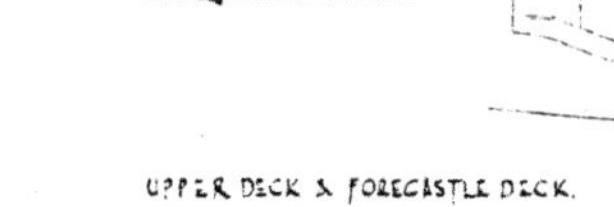

UPPER DECK & FORECASTLE DECK.

GUN TOMPION & BOAT BADGE

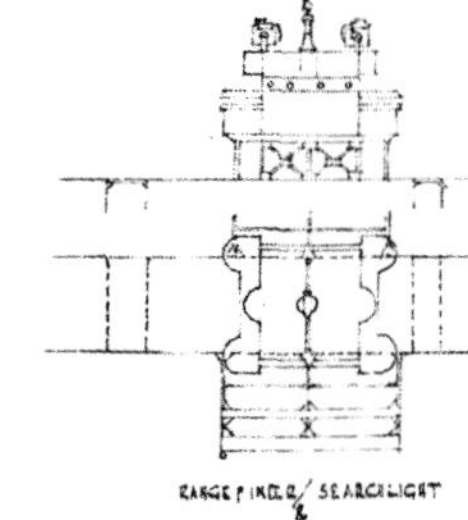

OTHER SHIPS

CORDELIA.	Pembroke Dockyard	July '13 - Jan. '15
COMOS.	Swann - Hunter Ltd.	Nov '13 - May '15
CARYSFORT.	Pembroke Dockyard.	Feb. '14 - June '15
CLEOPATRA.	Devonport "	Feb. '14 - June '15
CONQUEST.	Chatham "	May '14 - June '15

NO WAR LOSSES

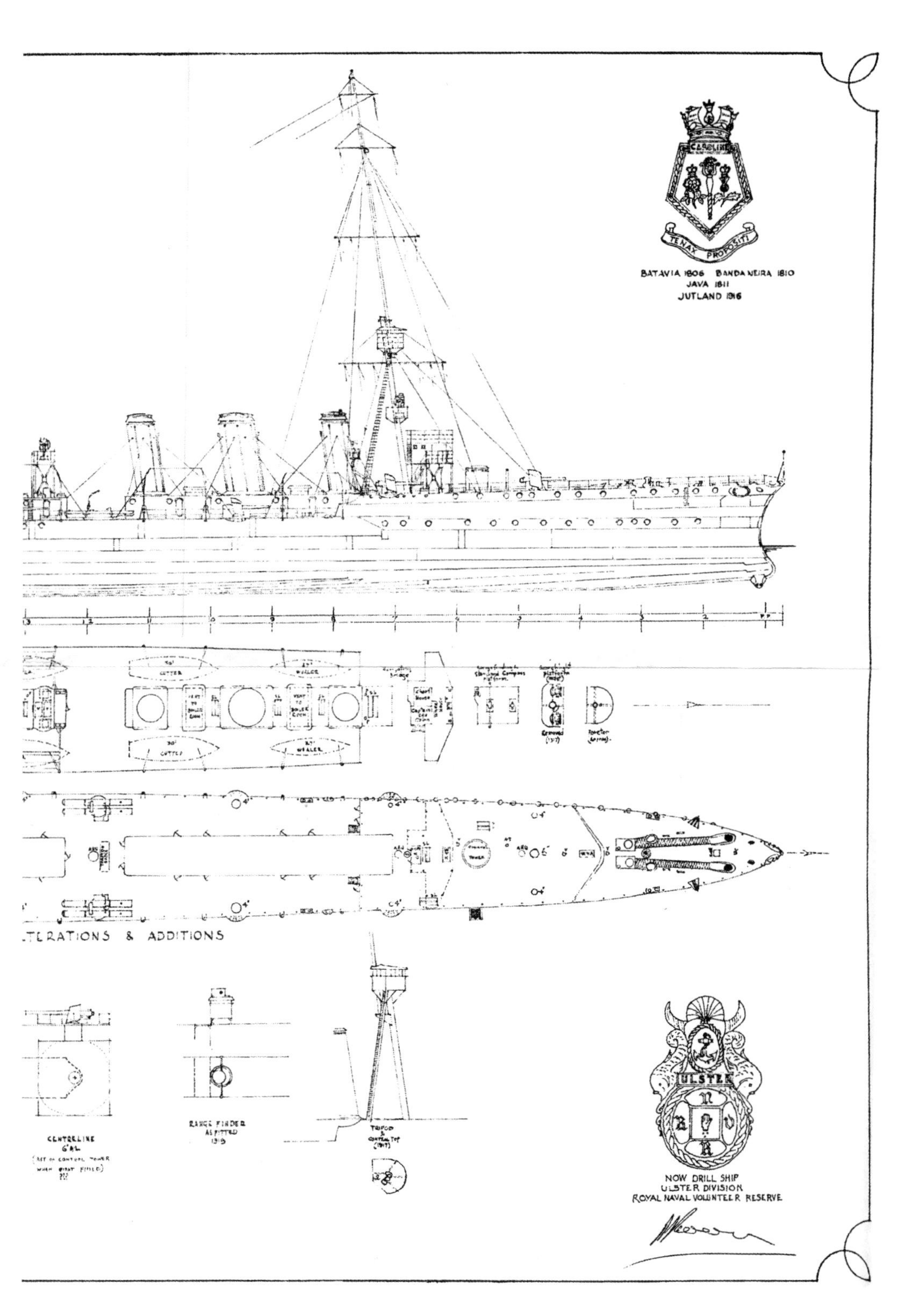

Plan of HMS *Caroline*. Behind, or abaft, of the rearmost funnel are her twin torpedo tubes, on port and starboard sides. The tubes where installed on a turntable to fire at a target at sea.

Author's Collection

HMS *Arethusa* at sea, name ship for the class of light cruisers that led to a refined and updated class of vessels – HMS *Caroline* and her sisters.

Public Domain

HMS *Caroline* undergoing full power sea trials off Birkenhead, December 1914.

Courtesy Wirral Archives Service

Leading steward, Albion Percy Smith who served in HMS *Caroline* from its commissioning.

Author's Collection

HMS *Iron Duke*, flagship of Admiral Sir John Jellicoe Commander-in-chief of the Grand Fleet of the British Navy.

Author's Collection

Admiral Sir John Rushworth Jellicoe, Commander-in-chief of the Grand Fleet of the British Navy.

Library of Congress, Prints & Photographs Division. Reproduction Number LC-DIG-ggbain-38732

Vice Admiral Sir David Beatty, commander of the Battle Cruisers of the British Grand Fleet.

Public Domain

Admiral Reinhard Scheer, Commander-in-chief of The High Seas Fleet of the German Navy.

Public Domain

Vice Admiral Franz Hipper, commander of the Battle Cruisers of the German High Seas Fleet.

Public Domain

A view of SMS *Blücher* sinking after receiving multiple hits from British ships at the Battle of Dogger Bank on 24 January 1915. This photo was taken from HMS *Arethusa.*

Public Domain

SMS *Von der Tann*

Library of Congress, Prints & Photographs Division. Reproduction Number LC-DIG-ggbain-16927

A panoramic photograph of battleships of the British Grand Fleet's 1st Battle Squadron in the North Sea, April 1915. Visible are HMS *Marlborough* (2R), HMS *Colossus* (3L) and HMS *Hercules* (4L). Other visible ships are HMS *Superb*, HMS *St Vincent*, HMS *Collingwood* and HMS *Vanguard*.

US Naval History and Heritage Command Photograph NH 2714

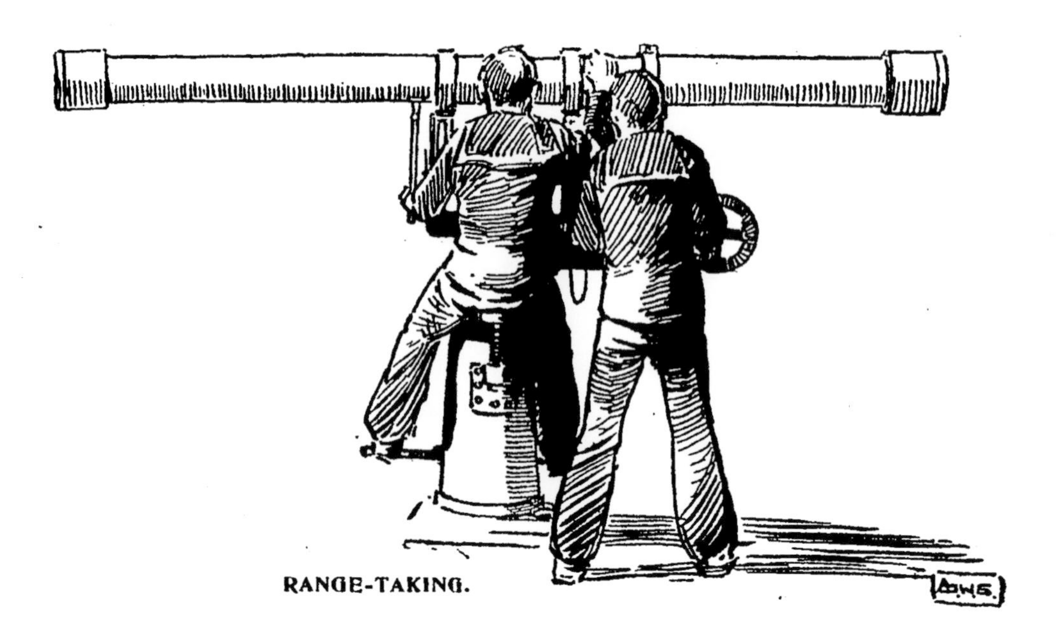

An illustration of range-taking, the role in which Albion Smith served during action stations in HMS *Caroline*.

Public Domain

Picture of a torpedo entering the water, fired from above the waterline.

Author's Collection

HMS *Indefatigable,* viewed from the stern, prior to its loss at the Battle of Jutland.

Public Domain

The terrible moment that HMS *Invincible* explodes only minutes after the 3rd Battle Cruiser Squadron joined battle.

Public Domain

HMS *Royalist*, a ship of the Arethusa Class (*Caroline's* predecessor) and companion in the 4th Light Cruiser Squadron during the Battle of Jutland.

A profile view of HMS *Caroline,* showing replacement tripod foremast in 1917.

Public Domain

Monday 1 February 1915 – To Rosyth

AM *Ship zigzagging every 19 mins*

PM *6.35 entered buoyed channel S 89° W*
7.10 eased to pass outer boom
7.15 passed under Forth Bridge

Tuesday 2, Wednesday 3 February 1915 – At Rosyth

Joined ship from Princess Royal, *2 ord. Seamen, 1 boy telegraphist, joined ship 2 armourer ratings*

Thursday 4 February 1915 – At Rosyth

PM *4.00 exercised 'repel aerial attack'*

Friday 5 February 1915 – At Rosyth

General quarters, collision stations, abandon ship station

Saturday 6 February – At Rosyth

Read Warrant No. 5, exercised fire stations.

From 7 February 1915 for the next nine days more evolutions were practised: marines at infantry drill, guns crews' first aid instruction, the hands turning out boats, burning search lights, divers down inspecting and cleaning inlets and propellers.

Periodically then, in 1915, *Caroline* crossed the North Sea with other light cruisers and destroyers to patrol the area off the Norwegian coast in the vicinity of the island of Utsire[1], stopping merchant ships suspected of carrying contraband. Indeed for the remainder of 1915 *Caroline* continued her sweeps of the North Sea as a zealous workhorse of Admiral Jellicoe's Grand Fleet.

There were, naturally, lighter moments: sing songs, concerts and amateur dramatics in which Albion Smith immersed himself. Clearly from his diary entries he had some musical aptitude, making a note of the titles of popular and favourite

1 Sparsely inhabited island about 15 miles off Norwegian coast and familiar weather forecast area

music hall songs such as: 'I loved you more than I knew', 'How's your father', 'Whoops, let's do it again', 'Gilbert the Philbert' and 'Burlington Bertie':

I'm Burlington Bertie I rise at ten thirty
And saunter along like a toff.
I walk down the Strand with my gloves on my hand
Then I walk down again with them off

But an unexpected summons to duty could at any time interrupt moments of leisure. One much loathed odd job that fell to the lot of light cruisers such as *Caroline* and her sister *Cordelia*, was summed up by the letters: DNP – Dark Night Patrol. Every night when there was no moon it fell to a light cruiser and two destroyers to make a foray, whatever the weather, two hours before sunset eastward of May Island at the mouth of the Firth of Forth. The purpose of these patrols was to catch any fast, lone German mine-layer at work off the port in the dark hours. How these stirring words to this song must have resounded with gusto in *Caroline*'s wardroom on the eve of such occasions:

So, blow up your Giever![2]
Come fill up your flask!
You've had time for luncheon,
what more do ye ask?
So open the Outer Gate,
let us gae free,
The signal is flying, 'Light Cruisers to Sea'

Gratefully the light cruisers returned to port two hours after sunrise.

2 Winter clothing supplied by Messrs Gieves and Hawkes of Savile Row, London.

Part Two

STOP
STANDBY ENGINE
DEAD SLOW
DEAD SLOW
SLOW
AHEAD
FULL
FULL

14

'Der Tag'

By spring 1916 HMS *Caroline*'s crew, now familiarly known as the 'Carries', itched for action at sea. Yet after sixteen months of the ship in commission they had not had so much as a brief skirmish with any of the enemy's surface forces. In February 1916 a German airship, a Zeppelin, had identified her in the River Humber and reported inflicting severe damage on the new light cruiser[1]. But her ship's company had no knowledge of this incident. Indeed due to censorship, because of the need to keep the enemy ignorant of his successes, the Carries would not even learn of the demise of *Caroline*'s elder sister. In February the plucky *Arethusa*'s strikingly active career came to an abrupt end, blown up by a mine and declared a total loss.

Yet no such news reached Albion Smith. Instead he wrote on Tuesday 8 February 1916: "Wild stormy weather, heavy snowfall. Left Scapa Flow at 2.30pm and had rough and bitter cold trip, ship rolling terribly all night, decks all awash". On Wednesday 9 February they: "Arrived at the mouth of the Tyne at 8.30 and tied up to buoy outside – until 6.00pm, then we went into dock. Heard news about the Caroline being lost with all hands??? *[sic]* Left ship to go on leave. Left Newcastle for London at 11.55. Train crowded with soldiers. Reached Kings Cross at 6.00am. Went to the Y.M.C.A. and to bed till 10.00am. Then went to GPO and sent wire to Chrissie[2]...went to her home at Ickenham..."

During Smith's leave they travelled to Brighton visiting friends, but next day Tuesday 15 February... "lovely morning on the pier and front. Beautiful sunshine. Wire came for me to return immediately. Left Brighton about 4 pm. Had a rather sad trip up to Vic' as both of us thought of the parting. On reaching London had tea and went on to KX[3], met others from *Caroline*. Said goodbye to Chrissie and started North once more."

1 "On the Humber a cruiser was hit and heavily damaged, she turned out to be, later received information [confirmed], the new small cruiser 'Caroline' (3810 tons)". *Germany's naval fleet in the World War: Personal Memories* by Admiral R Scheer, 1920.

2 Albion Smith's girlfriend at the time.

3 Kings Cross Station

On Saturday 26 February Albion Smith writes: "Left Scapa Flow about 4.30am. Out with the Grand Fleet on P.Z's.[4] The weather was most glorious. *Iron Duke*, *Canada* etc about 70 ships all told including destroyers. They made a grand picture spread out in battle order.'

Next day Smith records that they: "...joined up with Beatty's crowd about 100 ships all told... snow falling in afternoon, ship rolling about a great deal".

Although he could be a 'holy terror' to his crew in the words of Albion Smith, Captain Crooke was well liked. As we have seen, Crooke had been commanding officer of the Royal Navy's gunnery school at Whale Island, Portsmouth. *Caroline* was his first seagoing command thanks to a personal recommendation from Admiral of the Fleet Sir John Jellicoe. Crooke's next senior in rank was the Flag Officer of the Fourth Light Cruiser Squadron Commodore Le Mesurier in *Calliope.* Three light cruisers made up the squadron, including *Caroline* and *Comus.*

On Tuesday 25 April a surprise attack on Lowestoft and Great Yarmouth by German battle cruisers had taken place. Their battle fleet gave distant back up but soon returned to port. Once again Jellicoe's and Beatty's attempts to intercept German hit and run raiders had been foiled. As Albion Smith says: "Heard by signal about Germans bombarding Lowestoft. So tried to cut them off, speed about 24 knots... plenty of excitement and expectations but no luck... cold and wet with sleet and rain in early morning. Everybody disappointed at our bad luck."

Day in day out, month in month out had found *Caroline* on duty as a fleet work horse. When not at sea patrolling off the coast of Norway intercepting neutral shipping for contraband, and contributing to the distant blockade against Germany, *Caroline* carried out repetitive training drills either at Scapa Flow, Invergordon or Rosyth. These drills consisted of route marches, weighing anchor by hand, collision stations, fire and abandon ship stations, search-light drill, torpedo firing practice, and sub-calibre armament practice.

Of the three locations on the Scottish east coast a favourite was Rosyth where football offered much needed recreation, together with concert parties. Boxing tournaments and boat races were also popular.

Albion Smith, as we have seen, had long settled into the trusted position of captain's steward. He had been accorded a rating equivalent to Leading Seaman. Brought up in Rotherfield, his mother was a widow. He was one of a family of eight. The Lady of the Manor, Katherine Pullein, observed that Albion was a good scholar at school. In 1911 Miss Katherine Pullein recommended he join the Merchant Navy and assisted with his expenses. Smith's conscientious personality plus three years of sea experience in the Merchant Navy assured him of a place on *Caroline*'s bridge, as

4 Fleet tactical training exercises at sea

6 INCH GUN

A 6 inch gun's crew in action numbered nine men:

Gunlayer: Lays the gun for elevation

Trainer: Trains it for direction

Sightsetter: Adjusts the sights for range and deflection

Breechworker: Role is to open and close the breech

The remaining five of the crew were the loading numbers.

On the order "Commence", a projectile is rammed home and the cordite charge pushed in after it. The breechworker slams shut the breech and calls out "Ready". When 'the sights are on' the gunlayer presses the trigger. And with a blinding yellow flash and a roar the gun fires and recoils. At once the breechworker seizes the breech-lever to open it. With smoke still issuing from it another shell and charge are rammed home.

Illustration of a 6 inch gun crew.

his action station. There he was useful as a look-out and as a range-taker.

Off duty he wrote up his day in pencil in a small Letts diary. Now a century later it allows a fascinating comparison to be made between Smith's observations and the ship's official log at the National Archives Kew.

As spring 1916 arrived the question was, how could the entire German Fleet be brought to action at long last? After twenty months of hostilities, the Grand Fleet's patience was wearing thin. Would they *ever* come out, wondered Admirals Jellicoe and Beatty?

GERMAN REASONING

In Germany the mood was that something had to be done to relieve pressure on her army. The German army was at a standstill at Verdun. Therefore, the High Seas Fleet led by Admirals Reinhard Scheer and Franz Hipper, outnumbered as it was, decided on a date in May to win an equaliser, i.e. by destroying isolated elements of Jellicoe's squadrons to achieve parity with the Grand Fleet. But an all-out fleet encounter was to be avoided.

Admiral Scheer had become the supreme commander of the German fleet from the start of 1916. He saw how his enemy's battle cruisers had failed to stop the German hit and run raids on the English east coast. The English *Panzerkreuzern*, battlecruisers, were handicapped by the distance they had to travel from their bases in the north. Quite simply the English were not advantageously placed to give chase. Could not Hipper's force of five battle cruisers assist them to get closer? Trail their coats off the Norwegian coast to provoke the English to give chase?

And could not the enemy be led into a trap across the North Sea, ideally off the Danish mainland known as Jutland or in the vicinity of the Skagerrak? Moreover, would it not be there that the complete German battle fleet would lie in wait to spring the trap? The English striking force would be destroyed outright. German mastery of the North Sea and English Channel would follow. Britain's army on the continent would be stranded. The war would soon swing decisively in Germany's favour. Yes, the time had come to stand up to the sons of Nelson: "Denken wir nur an Auesserungen wie: [To think of those pretentious utterances]: 'We have the ships, we have the men, we have the money too' oder auch an Schiffsnamen wie [or the manner of naming their ships] *Irresistible, Invincible, Indomitable, Formidable...*'[5]

The enemy would not have things all his own way. German warships were stronger, more robust. Defects in magazine and shell handling compartments had been put right after the Dogger Bank battle in which an explosion had very nearly sunk the *Seydlitz*. German optical instruments, range finding appliances were second to none in quality, as was the quality of German munitions, and armour piercing shells. Through the German attaché at the embassy in Stockholm, Scheer had been told amusing stories of how English armour piercing projectiles had failed to penetrate German armour when struck at an angle. Their shells broke up, shattered into pieces. And in the unlikely event that he, Scheer, may find his fleet confronted unexpectedly by the entire English battle fleet, then they would put into practice their well rehearsed: *Gefechtsabkehrung* – a 'battle turn away'. Scheer and

5 *Germany's naval fleet in the World War: Personal Memories* by Admiral R Scheer, 1920

Hipper anticipated 'Der Tag' with a new found confidence. The 15 May was to be '*Der* Tag' – *The* Day.

Meanwhile German submarines took station off Grand Fleet bases. Zeppelins burbled menacingly at cloud level to track English fleet movements and to bomb towns on England's mainland.

So, it was that the confined arena of the North Sea was about to play host to the greatest sea battle of the whole war. The 'Carries' had no inkling of an impending fleet action. Departing Scapa was to them just another movement order; a routine patrol, as can be seen from these selected log extracts from early May:

Monday 1 May 1916 – At Scapa

AM *Single anchor wind various, force 0–1 temperature 58*
6.00 hands employed cleaning ship, preparing for drill
9.00 divisions, prayers
9.30 let go second anchor, away all boats pulling round the squadron, weigh by hand
10.00 hoisted all boats.
11.00 lit fires in A3 A4 B3

PM *1.00 lit fires in B1 B2 B3*
1.10 weighed, proceeded, carried out gunnery and torpedo practices in Flow course and speed as req.
4.15 let fires out in A3 A4
4.20 finished practices picked up torpedoes and returned to anchorage
5.00 stopped came to starbd anchor 6 shackles, 17 fms in Y4 berth
5.15 let fires out in B1 B2 B3

Albion Smith writes: "Beautiful sunshine. Left anchorage at 1.30pm to go out in the Bay firing sub-calibre and torpedoes. Returned to our billet at 5.30pm. Free gift of vegetables and jam. Most glorious sunset".

Boiler room ratings, known as 'stokers', tending *Caroline*'s eight Yarrow boilers, very likely blessed the day they received a draft to *Caroline*. No shovelling coal into blazing furnaces, no need for 'trimmers' sweating away in the grimy confines of stokeholds and coal bunkers. *Caroline* raised high pressure superheated steam

by her eight Yarrow built oil-fired boilers to drive a group of Parsons[6] turbines, embedded in two separate engine rooms. They delivered 36,000 shaft horsepower to four propellers.

Tuesday 2 May 1916 – At Scapa, Wind easterly 1.2

AM *6.00 hands employed cleaning ship*
8.00 exercised physical drill
9.00 training classes at instructions. Landed marines for drill
11.00 guns' crews at divisional drill
12.00 paid monthly advance

PM *Landed football parties*
6.00 secured for sea, tested life buoys found correct…

Albion Smith writes: "Fine but dull morning. Pay day £1 service pay. Gave steward 10/- on account for suit. Rain came on about 4 pm"

Wednesday 3 May 1916 – At Cruising

AM *Calm, 745 tons fuel oil*
0.25 lit fires in B1 B2 B3
1.12 lit fires in A3 A4
2.45 weighed
3.30 proceeded as req, 3.50 ahead 15 kts
4.03 passed outer boom
4.20 14 kts, 4.40 18 kts 246 revs
4.50 formed single line abreast to port 5 c apart S 82° E 15 kts 276 revs
5.15 zigzagged 3 pts[7] each way every 10 minutes 14 kts
7.30 15 kts,
7.45 a/c S 78° E

PM *12.00 Pat. Log 105.7, 277 revs, wind SSE force 3, Barometer 29.65 temperature 53*

6 Named after the inventor of the turbine: Sir Charles Parsons
7 Pt: A point, i.e. 11¼ degrees (32 × 11¼ = 360 degrees)

1.00 Pat Log 127.1, single line ahead close order in station co. S 48° E, 16 kts
1.40 take up position spread from Calliope *10 miles apart, 24 knots*
3.20 S 50° E
3.35 a/c as req to starbd to keep in visual touch with Royalist
3.55 passed trawler BL 84 Holland steering north, wind south, force 3
7.35 incd 20 kts, a/c N 50° E
7.47 20 kts for taking station on Calliope…

Albion Smith writes: "Left Scapa at or about 3.30. Very dull morning with slight mist and rain. All the Grand Fleet out, preparing for an attack on Heligoland. Passed several neutral ships. A little linnet flew about the ship."

Once beyond the outer boom at Scapa the squadron increased speed to 14 knots. Tense sharp-eyed lookouts on the ship's open bridge scanned the sea for tell-tale traces of lurking enemy periscopes. So the ships zigzagged. For defence a warship would try to ram a U-boat and zigzag every five minutes at speed. The depth charge/bomb was at this time still a year away. Note after forming single line abreast the squadron kept a distance of 5 cables apart from each other.

Thursday 4 May 1916 – At Cruising

AM *Pat. Log 331.9, wind west force 2*
3.05 a/c S 25° E
3.30 18 kts
3.35 a/c S 32° E
3.40 S by E formed line ahead to port, 7 c, log 377 miles
4.30 action stations
6.20 14 kts
6.25 took station on port wing of B.F. [battle fleet]
7.00 Abdiel[8] *re-joined, passed close to* Heemskirk, *Dutch trawler with trawl down Lat 56° 29 minutes north, Long 5° 16 minutes east,*
7 .15 moved to bearing NNW from Flag
7.45 a/c NNW took station 4 miles ahead 16 kts
8.00 zigzag 1 pt each way
8.15 line abreast to starbd 7½ cables intervals

8 HMS *Abdiel* a destroyer-minelayer, 1,687 tons with capacity for 70 mines.

8.30 passed close to small iron buoy

10.05 sighted steamer NE 10 miles steering easterly.

10.05 passed close to Flora, *Denmark steamer steering E by S 56° 49 minutes North , 5° 43 minutes East*

10.30 wind SW, 2nd Div 3 miles from Flag Co SE by S, zigzag 1 pt each way, 17 kts

PM *Pat. Log 510.5 miles*

0.20 In station s/c[9] *S by W 16 kts, spd as req for keeping station.*

2.00 hands at day cruising stations

3.10 co. NW by W 16 kts

3.15 let fires out A3 A4

4.10 negative zigzag

5.17 dead slow formed S.L. [single line] ahead

5.20 W by N 10 kts

5.25 14 kts. Pat Log: 669 miles

Albion Smith writes: "Dull grey morning. All hands out for action stations at 4.30. At general quarters all the morning, passed the Danish steamer *Flora* and others name not known. Heard the sound of gunfire about 12 o/c and in the afternoon we were told that a Zeppelin had been brought down. All of us disappointed at the fact that it wasn't our capture. Very dense fog came on."

Friday 5 May – To Scapa

AM *11.46 Took station on* Calliope *S.L. ahead, co & spd as req for entering harbour*

PM *Y4 berth 869 miles*

Albion Smith writes: "Dull morning but sea still calm. Reached Scapa about 12.30. The ships coming in thro' the entrance were in a line over 15 miles in length. Rain came on as we anchored…"

9 s/c: set course

Saturday 6 May – At Scapa

AM *9.00 mustered at divisions with dirty hammocks*
10.00 hands employed clg ship

PM *Make and mend clothes.*[10]

Albion Smith writes: "Very wet morning. Busy cleaning up all day."

Sunday 7 May – At Scapa

AM *9.00 performed divine service*
11.00 read articles of war
11.30 discharged to shore canteen manager

PM *2.10 lost overboard by accident from skiff*[11] *crutches GM Patt 73e one in No. [probably timber supports for the skiffs mast]*

Albion Smith writes: "Fine morning, a little sunshine."

Monday 8 May – At Scapa

Exd [exercised] search light control

Albion Smith writes: "pouring in torrents of rain. Went out in the Bay taking ranges. Rain eased up about 12 but it was showery in the afternoon."

As German U-boats took up their stations to attack warships either entering or leaving harbour ahead of the 15 May, *Caroline* prepared to join another North Sea sweep.

Tuesday 9 May 1916 – At Cruising

AM *Draught fwd: 15 ft 3 in, Aft: 16 ft 1 in, 965 tons oil fuel*
5.15 lit fires A2 A3 A4, employed cleaning ship, securing for sea

10 A half holiday, usually a Thursday, for the ship's company to repair and replace their kit.
11 Skiff: one of the ship's boats

7.45 weighed, 7.55 procd 12 kts co. as req
8.14 passed outer boom
8.55 s/c N by E to spread 40 miles NNE ½ degree E from Calliope, *22 kts.*
9.40 streamed patent log
10.30 a/c N 83° E

Albion Smith writes: "Showery morning, up at 5.15 am. Left S Flow at about 7.30 to go on patrol work off Norway. Rain all day. Passed Swedish ship *Mimosa*, 965 tons, cargo of wood."

Wednesday 10 May 1916 – At Cruising

AM *0.45 passed SS* Nieuw Amsterdam *steering N by W*
1.56 a/c S 76° E
2.15 a/c to examine steamer Annam, *Copenhagen to New York*
2.25 resume co. S 76° E
2.55 sighted Udsire lt. on port bow bearing east 3 miles, a/c as req to proceed down Norwegian coast keeping 3.4 miles from shore
5.25 examined SS Parvus, *Stockholm*
7.10 stopped and boarded SS Albania, *Swede*
7.15 sighted Calliope
7.40 picked up boarding boat
7.45 proceeded as req to join up with Calliope...

Albion Smith writes: "all hands out at 4.00am to general quarters. In sight of coast of Norway. Very dull morning, sea very calm. Held up Swedish steamer *Parvus*, 1692 tons at 5.25 am but did not board her owing to her being inside 3 mile limit neutral waters. At 7 to 7.30 we held up *Albania* a Swedish steamer, 541 tons and boarded her after firing a warning but nothing doing."

15

TORPEDO TANTRUMS

Thursday 11 May 1916 – To Rosyth

AM *1.00 sighted May Isle lt*[1] *SW*
1.05 a/c S 52° W
1.35 red 15 kts, 1.40 a/c S 79° W
2.00 passed gate, passed bridge

Albion Smith writes: "Reached F of F at 4.00am about. Dull morning. Mrs S[2] came to lunch and tea."

FOR THE CARRIES it was surely a welcome break from the dull routine of the usual North Sea sweep to enter the Firth of Forth for thirteen days, dock the ship in Rosyth and make repairs to her torpedo boxes. To facilitate reloading torpedoes while at sea two splinter-proof containers for the storage of spare torpedoes were installed next to the launch tubes. A defective 6 inch gun also needed repair.

It may be noted that Captain Crooke was stepping out at this time with the widow of a Lieutenant Colonel of artillery. Mrs S also came to lunch on Friday 12. On May 13 Smith records the captain as being "away to lunch but came aboard to tea with Mrs S. Capt. went out to dinner. Lovely moonlight night."

On Monday 15 May setbacks hit the proposed German sortie. It happened that the U-boats already lying in wait had run out of fuel by now. They were withdrawn. 'Der Tag' was put off until Wednesday 31 May. This extra time was needed to repair mine damage to Admiral Hipper's newest battle cruiser, *Seydlitz* to bring his force up to the acceptable minimum of five[3].

Albion Smith writes: "Beautiful morning, came down into Rosyth dock about midday. Steward and I went to Edinburgh about 5.30pm. We visited the castle and

1 May Island light
2 Mrs Lilian Ethel Smith
3 *Von der Tann, Moltke, Lützow, Derfflinger, Seydlitz*

had a grand view of the town and saw most of the places of interest. The city is beautiful. We went to Kings Theatre and saw Wilkie Bard. After this went and slept at Buchanen's Hotel..."

On 16 May, Smith records that he left Edinburgh at 6.30am and got to the ship 7.45am… "dull morning but cleared up, later sun shone. Capt. had Mrs S to lunch then went ashore to tea."

On 17 May Smith wrote: "Nasty wet morning. Capt. sent me to Elgin Hotel Charlestown to bring off suitcase. Went up by motor boat. Ship moved out of dock by midday and we went upstream and anchored near *Calliope*. Mrs S to lunch and tea. Capt. had dinner in wardroom."

Monday 15 May 1916 – At wet basin Rosyth

AM *05.30 hands employed cleaning ship*
07.00 prepared for general drills
09.00 exd lay out bower anchor – away all boats crews, weighed by hand
10.00 Relief boats crews away, out all life saving rafts
11.00 hoisted all boats, stowed all ammunition below
11.15 weighed, towed by tugs to Rosyth dockyard

PM *0.40 made fast alongside wall in basin*
1.30 party landed for recreation, remainder assisting with torpedo boxes, dismounting Z2 6 inch gun, attending floating crane, and as req.
6.00 watch employed working about torpedo boxes, 6 inch gun and as req

---

THE 21 INCH TORPEDO

Caroline possessed the very latest and most destructive missile on offer in Britain's naval arsenal in 1914; the 21" torpedo. Twenty-one inches in diameter and twenty-four feet long. In the front is the explosive head of gun cotton with percussion fuse. The main body of the cylinder contained compressed air. Behind the air chamber lay a compact engine. In the earliest torpedoes the compressed air supply was soon exhausted. However by 1911 a heater was installed to enhance power and speed. A gyro appliance within the body of the torpedo adjusted the missile's balance and directed its course as required. The introduction of air-heating enabled torpedoes to travel 1000 yards at 40 knots, 2000 yards at 37 knots and 4000 yards at 27 knots.

Tuesday 16 May 1916 – At wet basin Rosyth

AM *09.00 hands employed painting side, party cleaning, lubricating aft 6 inch gun*

PM *Leave to chief and P.O.s[4] till 6.30pm*

Wednesday 17 May 1916 – At wet basin Rosyth

AM *06.00 hands employed cleaning ship, party painting side*
09.00 hands painting side, painting torpedo boxes
10.15 lit fires in B1 B2 B4. Number on sick list: 1

PM *00.25 moved from jetty into lock*
00.35 proceeded out of lock
1.03 stopped
1.05 came to stbd anchor – 7 fms in B6 berth
1.30 let fires out in B1 B2 B4, party landed to attend concert, remainder employed getting up ammunition as req.

Thursday 18 May 1916

AM *6.00 hands employed cleaning, boats etc*
8.30 aired bedding
9.00 training classes at instructions, remainder painting torpedo boxes, side as req.

PM *1.00 make and mend clothes*
2.00 leave to Chief and P.Os
5.00 read warrants 48, 49, 50. Watch employed getting up torpedoes
9.00 exd watch at search light control
9.30 shortened in to 3 shackles.

Albion Smith writes: "lovely morning. Warm sunshine, quite a heat wave.

4 Petty Officers; non commissioned officers

Mrs S + 4 others to lunch. Mrs S stayed to tea. Busy time... in evening swung very near the Calliope."

The length of a shackle of anchor chain cable was measured in fathoms –12½ fathoms, 75 feet. As Smith observes, *Caroline* was swinging too close to the Flagship. So an amount of *Caroline*'s chain cable lying on the bottom was heaved in or shortened to restore her to her allotted berth.

The next day's routine called for: 'stations for abandon ship' and 'marines at gun drill'. Part of the ship's company was landed for a concert at Rosyth yard. Meanwhile Albion Smith was kept busy: "Mrs S and Captain Townsend to lunch, Mrs S to tea... free gift of fruit and vegt to the ship... thick fog again at night.' Sunday 21st slight fog in the morning but sun came out after. Mrs S & sky pilot to lunch and Mrs S to tea. I went ashore to Dun [Dunfermline?] and saw Scots regt. on their way to station for Egypt. Came on to rain and came back in the wrong bus and had to run to reach the ship in time. Got rather wet and tired.

On Tuesday 23 May Albion Smith writes: "Dull morning, inclined to rain. Mrs S came to lunch. Two sisters + sky pilot[5] and Mrs S to tea."

Wednesday 24 May 1916 – Cruising from Rosyth

AM *0.15 lit fires in A1 A2 B1 B2 B4*
2.50 weighed
3.20 passed bridge
3.40 passed boom, 18 kts co. as req
4.30 passed 2nd LCS
5.00 Royalist, Acheron, Ariel *in company*
6.25 sighted Brit. Seaplane...

Albion Smith writes: "Left FoF *[sic]* at 2.30am in company of other light cruisers and destroyers. Fine morning but the air was a bit cold. G quarters in the forenoon. Uneventful day."

From these brief diary entries, as spring awakens and warmer days beckon, a sense of frustration at the lack of action with the enemy plainly besets Smith and the ship's company. Perhaps they now felt a moment of glory for their ship was at hand – but when?

5 Possibly in charge of observation balloon or a 'waterplane'.

Thursday 25 May Albion Smith writes: "Lovely morning. All hands at action stations at 4.00am. Off the coast of Norway the country looked very pretty. Norwegian destroyer followed us for a long time to keep us from infringing neutral waters. Held up a steamer and passed Swedish fishing fleet..."

On Friday 26 May *Caroline*: "Reached Scapa Flow about 2.30am. Thick fog but it cleared later and became a fine day. Had G quarters at 10.00am... saw observation balloon right overhead..."

Saturday 27 May: "Glorious morning. Busy cleaning all day. *Canada*'s[6] band came aboard in the evening. Lovely sunset." And on Sunday 28 May 1916 Albion Smith writes: "Fine morning and turned out a grand day with splendid sunshine..."

6 HMS *Canada* ,26,000 tons, dreadnought, completed 1913

16

Raise Steam with All Despatch... The Eve of Battle

Monday 29 May 1916 was an uneventful day as borne out by Smith's usual reporting style: "Dull morning but it became a glorious sunshiny day. Ship went across to the north shore near Kirkwall to prepare for firing. Lovely sunset in evening." Then, whether it was an inspired foretaste of what was to erupt on the morrow or mere happenstance we shall never know, but the next day was selected for extensive torpedo trials as recorded in the log.

Tuesday 30 May 1916 – From Scapa to Sea cruising

AM *06.00 hands employed cleaning ship*
06.30 torpedo ratings preparing torpedoes for running
09.00 exercised physical drill, employed painting gunshields as required
10.00 training classes at instructions
10.03 classes at .303 aiming rifle practice
11.00 boys at school
11.30 lit fires in A2 A3 A4 blrs. Wind SSW force 2 oc [overcast], sick list 1, started with 820 tons fuel oil, draught fwd 15 ft 0 in aft 16 ft 7 in

PM *1.50 weighed [anchor] and proceeded into Flow for firing torpedoes – carrying out torpedo box trials*
3.30 let fires out in A1 A2 blrs
3.40 finished commenced swinging ship for deviation landing compass
6.10 finished swinging, hoisted all boats, secured for sea
7.10 proceeded as req for anchoring in Fleet anchorage
7.40 stopped came to port anchor, 5 shackles 18 fathoms y4 berth

At this moment – in the aftermath of a good day at torpedo trials, following the repairs at Rosyth, the 'Carries' very probably expected an early night in. We see from Smith's diary that Captain Crooke invited his torpedo lieutenant (Lt. Ingham) to lunch: "Fine morning again. Carried out torpedo and gun firing while steaming around the bay. (Torp. Lieut. came to lunch). Went back to our old anchorage at 7.30pm. Another glorious sunset..."

On this day a certain room in Whitehall in the red brick new Admiralty building that faces Horse Guards Parade received intelligence from intercepted German wireless messages to the effect that the entire German *Hochseeflotte*, the High Seas Fleet, was raising steam. To Room 40[1] it could only mean one thing. The Hun was stirring. He was preparing for sea. He was at long last about to come out. Room 40 staff recognized the Morse characteristics of German ships' call signs. They passed their information to Admiralty planning staff in the same building. Soon a signal on this news was sent by wireless telegraphy to the Commander-in-Chief of the Grand Fleet, Admiral Sir John Rushworth Jellicoe. At this in turn the Grand Fleet was put on alert by visual signal at Scapa, Cromarty Firth (Invergordon) and Rosyth; a single flag was hoisted to the masts' yardarms of *Iron Duke, King George V* and *Lion* – a summons to: 'Raise steam with all despatch and report when you are ready to proceed.'

The matter of getting ready to put to sea in a hurry could be somewhat fraught in the sense of a controlled panic. It was not a case of turning a key and pressing a starter button. Both coal burning and oil burning ships needed time to heat fresh water and make steam to the required pressure. *Caroline*'s capacity for fresh water was in the region of one hundred and eight tons. We have seen from *Caroline*'s log how at times merely two or three of her eight boilers were fired up, for example: 'lit fires in A1 A2 and B3'. In theory a warship was entitled to claim four hours notice to raise steam from cold boilers to 'full pressure' if a signal came. The reality of life was, however, that a ship's engineers would be red faced with embarrassment against the better times achieved in colleagues' ships if their boilers were not ready within *two* hours.

The scene in *Caroline* may be imagined as she returned to her usual berth in Scapa. There might not have been time to send libertymen ashore (all her boats had been hoisted in). Officers sat comfortably in the wardroom playing cards, talking shop, arguing; a mood of dormant ambiance suddenly interrupted by the arrival of

1 A top secret part of the Admiralty Operations Division that comprised a team of experts using wireless telegraphy to decode German signals. They could tell which warships were in harbour and/or which were at sea by respective ships' call signs – much facilitated by the enemy not keeping wireless silence.

a boy with a cipher signal for the captain's secretary. Frowning, the secretary would rapidly vanish to decipher it. Someone would ask: "Is it a flap?"[2] A very natural query, until the secretary would duly return with a stern faced private confidential air, for only he and senior officers at this stage needed to know the signal's import. As others held their breath he would turn to the executive officer, to whisper in Lt Cdr Drummond's ear. Only when the executive officer whispered to the engineer commander DE Juke would everyone know a flap was on.

Watch keeping officers would now need to sort out who took the first and middle watches. The surgeon lieutenant, EF Murray, MD and sick berth attendants would hasten to place their medical kits at pre-selected dressing posts around the ship. Stokers on their mess-deck would hurry below to the fore and aft boiler rooms. Special sea duty men would leave their mess and hasten up the hatch to fall in on deck. Electrical ratings would hasten to fix shaded blue lights. The fo'c's'cle deck officer would get his party ready to weigh anchor.

Ready-use ammunition and cartridges had to be got up. Guard rails had to be taken down. The accommodation ladder had to be got in. Ship's mail for the shore probably could be sent over to the odd ship that was not going out. The clanging of bells would be heard as the engine room ratings tested the Chadburn steering engine and the telegraphs – the stand alone instruments that pointed to 'stop', 'slow ahead', 'half ahead', 'full ahead', 'astern' and indicated engine revolutions.

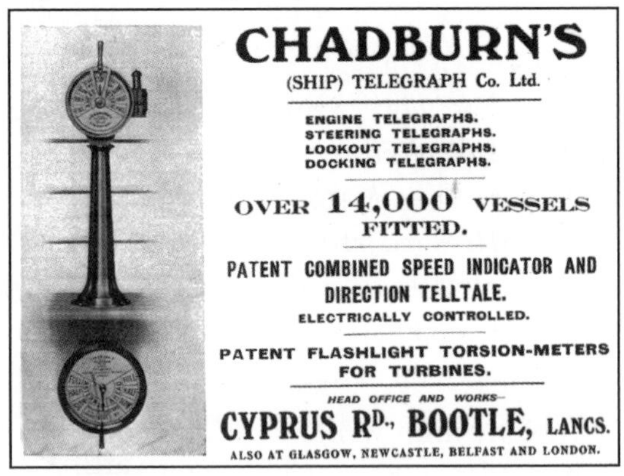

Advert appearing in The Shipbuilder Monthly.

Caroline had anchored in her usual berth at 7.40pm on Tuesday 30 May under a glorious sunset. Earlier, fires in A1 and A2 boilers had been allowed to die out by 3.30pm. However the signal arrived to raise steam again, and by 9.30pm *Caroline* was already able to signal: 'Ready to proceed'. It was acknowledged by her squadron flagship, *Calliope*. By now the fleet anchorage was wreathed in billows of black smoke as ship after ship was ready with steam. Of course everyone was well practised. They had been through this routine time and again. Even now at this stage they might still have been under the impression this was just another 'flap', unaware that in twenty-four hours time they would all be in the thick of a raging battle.

2 An alert that needed investigated but that would often prove to be a false alarm.

17

CAROLINE MEETS HER DESTINY – THE CLIMAX

Tuesday 30 May 1916

PM *9.45 weighed proceeded as req. for leaving Scapa*
10.08 passed boom outwards increased to 18 kts
10.15 telemotor failed, steered by engine and by hand courses as req, spd 20kts. Parted company with 4th LCS
11.03 passed Skerries N 5° E 2 miles, co. S 60° E 20 kts
11.08 a/c N 67° E
11.13 P. Skerries N 40° W 22 kts
11.28 red [reduced] 20 kts 385 revs
11.31 connected up steering engine gear s/c S 78° E. Sighted 4th LCS on stbd bow

---

Albion Smith writes: "Left Scapa for sea (with fleet) at 9.30. Steering gear broke down soon after. Got it in running order soon after 11.00pm."

THE CARRIES IN the course of working up and shaking down their ship had exercised many evolutions at sea to counter all manner of emergencies. Doubtless the evolution for 'steering breakdown' was practised. But it does not merit mention in the log. Anyhow, at a moment of crux on the eve of battle this was a serious emergency. Suddenly, as *Caroline* took her leave of the armed trawlers, the flotation buoys and the nets that formed the boom defences in Hoxa Sound, and slipped along at a racy eighteen knots, her means of power steering failed in waters where the tide can resemble a mill-race. The quartermaster in the wheelhouse would have reported losing steerageway. At this the officer of the watch would have told off seven men to man the emergency steering position right aft over the rudder to steer the ship literally by brute strength.

Meanwhile, *Caroline* fell out of line and steered as well as possible with her port and starboard propellers. In the 'tiller flat'[1] right aft in the stern of the ship six burly able seamen grasped the spokes of each of the three very large wooden wheels, two men per wheel. An extra hand stood by the voice pipe to acknowledge helm orders from the bridge and relay them to the men at the three wheels.

As *Caroline* regained her place with the 4th Light Cruiser Squadron, resuming course South 78° East, sixteen dreadnoughts under the direct command of Admiral Jellicoe emerged from Scapa Flow, keeping station with one another and in line ahead. Soon they were joined by Admiral Jerram's eight dreadnoughts rolling out of the Moray Firth. In the van of this extraordinary long line of twenty-four ships of the line was Admiral Hood with three battle cruisers, supported by the 4th Light Cruiser Squadron and destroyers. This combined force coming down from the north was to rendezvous at 2.30pm south of Norway (Wednesday 31 May 1916) with Admiral Beatty's six battle cruisers, supporting light cruisers and destroyers coming up from the south, out of their base Rosyth. There, in the vicinity of the entrance to the Baltic, the light cruisers were to try to bring the enemy to action.

This was truly a fleet of immense size with which to meet the less numerous German host making its way out of the Jade Estuary and up the Danish coastline with much the same objective in mind. It started north in the early hours of that day. Just in case his punch was not heavy enough to guarantee a knock out blow to the enemy, Admiral Jellicoe had extra back up – an especially powerful squadron. It was like a police force's special patrol group, a riot squad. To put it simply this group of super-dreadnoughts was akin to a roving trouble-shooter. Lying in Rosyth, this was the powerful 5th Battle Squadron; four ships mounting 15-inch guns, led by Admiral Evan-Thomas: *Warspite, Malaya, Valiant, Barham.* Their very names hinted at absolute dominance. The admiral's instructions were to support Vice Admiral Sir David Beatty. However, the two men had not worked with each other before this time. Would they harmonise for the crisis later in the day?

WEDNESDAY 31 MAY 1916

Albion Smith writes: "Dull but fine morning. Capt. had us all aft to tell us we were bound for Norway on a raiding trip after German cruisers. At 3.30pm we heard that Beatty was engaging German ships. Got called to action stations about 3.00pm…"

1 This cramped compartment deep below decks is entered via a steel hatch the size of a manhole cover. A vertical steel-runged ladder leads to a voice pipe, a rudder angle indicator and three outsize wooden wheels. Remarkably the fit out today in this compartment is as it was in 1916.

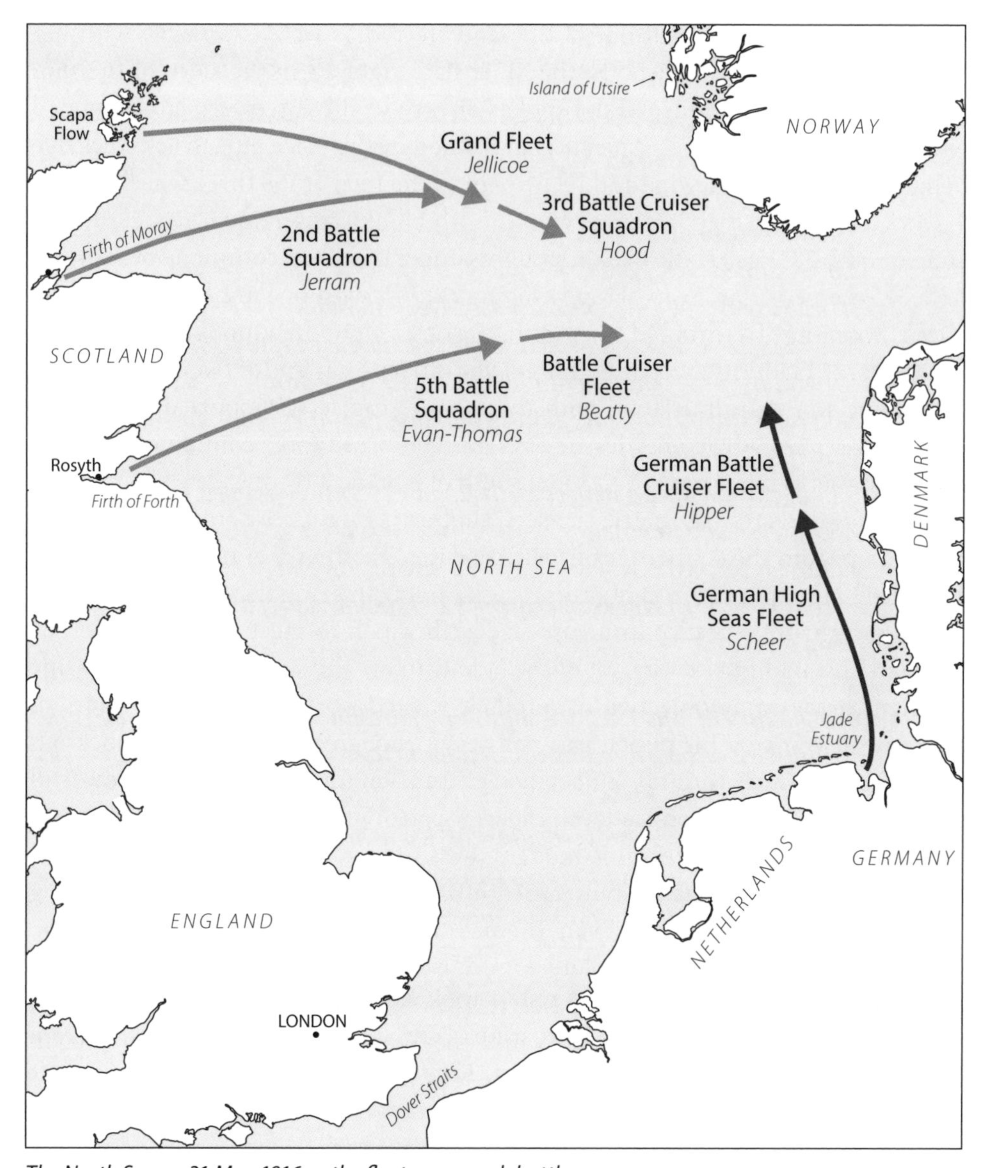

The North Sea on 31 May 1916 as the fleets approach battle.

Let us transcribe in full *Caroline*'s log, remembering that *Caroline* was in the van of the battle fleet coming down from the north, accompanied by her sister light cruisers and screening destroyers. The captain's address to the crew made it plain that this trip was no false alarm, they would return to Scapa with honour or not at all.

Wednesday 31 May 1916 – At Sea

AM *0.30 in station co S 78° E 18 kts*
1.30 red 17 kts, 1.50 a/c S 72° E
2.35 a/c N 10° E
2.45 turned 16 pts to stbd
3.00 resumed co S 72° E
3.12 15 kts zigzagged 2 pts each way every 10 miles
3.45 s/c S 72° E 17 kts. Formed single line abreast to stbd 7½ cables apart [with the other ships of 4th Light Cruiser Squadron]
4.00 co & spd as reqd for taking station on fleet [i.e. in the van as the eyes of the Grand Fleet]
5.13 in station ahead of fleet S 50° E, adv 15½ kts zigzagged 1 pt each way
7.50 streamed patent log
12.00[2] *lat 58° 031½ N, Long 3° 04 E, Norwegian Naze [A cape at the southern tip of Norway] bearing 92° 126 miles, 680 tons oil fuel, No. on sick list 2*

PM *00.30 2 iron drums Patt. 30 containing paraffin, 4 iron drums Patt 15 thrown overboard in preparation for action*
1.07 advance 13½ kts
1.38 adv 12½ kts. Lost overboard, carried away by K George V *rotators, [patent] log 1 in No.*
2.55 lit fires in B1 B2 blrs…

In the early hours, 2.45am, the ship took avoiding action, turning through 180°, possibly in response to a submarine scare? By the early afternoon there is some evidence of bunching or crowding, *King George V* being so close astern of *Caroline* as to run down her patent log. The general advance, all ships proceeding at equal speed, was slowed to twelve and a half knots. By 2.55pm word had reached *Caroline* that *Galatea*, one of Beatty's scouting light cruisers had signalled: 'Enemy In Sight'. *Galatea* had spotted Admiral Hipper's screening force of light cruisers and destroyers and, slipping along astern of them in all their majesty, Hipper's battle cruisers: *Lützow, Seydlitz, Moltke, Derfflinger* and *Von der Tann. Caroline*'s remaining boilers were lit.

2 The custom being that every ship of the fleet at sea signals her latitude and longitude according to her reckoning at midday.

GENERAL PROGRESS OF THE BATTLE

Phase one of the Battle of Jutland had started. This Battle Cruiser action is known as 'The Run to the South'. Beatty, who rode to hounds in his spare time ashore, now abandoned his northerly course to make the 2.30pm rendezvous with the battle fleet. Without waiting for the 5th Battle Squadron, which was some distance astern, Beatty cut to the chase going after the five German battle cruisers. In this action Beatty lost two of his ships: *Indefatigable* at 4.04pm (to *Von der Tann)* and *Queen Mary* at 4.26pm. They blew up. The four remaining British battle cruisers carried on their close engagement with the five German ships (at a range of 6 to 9 miles).

At 4.37pm a large force of German dreadnoughts was sighted tracking up from the southward – the High Seas Fleet in its entirety. At 4.45pm Beatty turned his ships away, one by one, right about, i.e. not together, northward. His ships received further punishment while executing this manoeuvre. The fierce duel, a dance of death, now continued on a northerly course, the German ships having also turned to the north.

Meanwhile the powerful 5th Battle Squadron had been following Beatty some four miles astern. They grouped up to the southward of Beatty's chastised force and took on the van of the German High Seas Fleet and the rear German battle cruisers. By 5.00pm Jellicoe was about 40 miles to the northward of this fast moving fight. He hurried south to join the battle as Beatty and Evan-Thomas led the German fleet on towards Jellicoe's trap.

To continue *Caroline*'s log: It records the afternoon's momentous action in some detail. Minute by minute one calamitous event follows another. Notwithstanding its understated prose there is a sense of ever-present mortal danger amid the ship's gyrations in this furious sea battle:

Wednesday 31 May 1916 [continued]

PM *3.00 a/c SE by S*
3.05 action stations
3.08 18 kts
3.26 19 kts
3.30 BCF[3] *in touch with enemy,*
4.00 log carried away, revs 337, wind various force 0–1
4.10 20 kts
4.15 neg z.z [ship not performing zigzag]

3 BCF: Battle Cruiser Fleet

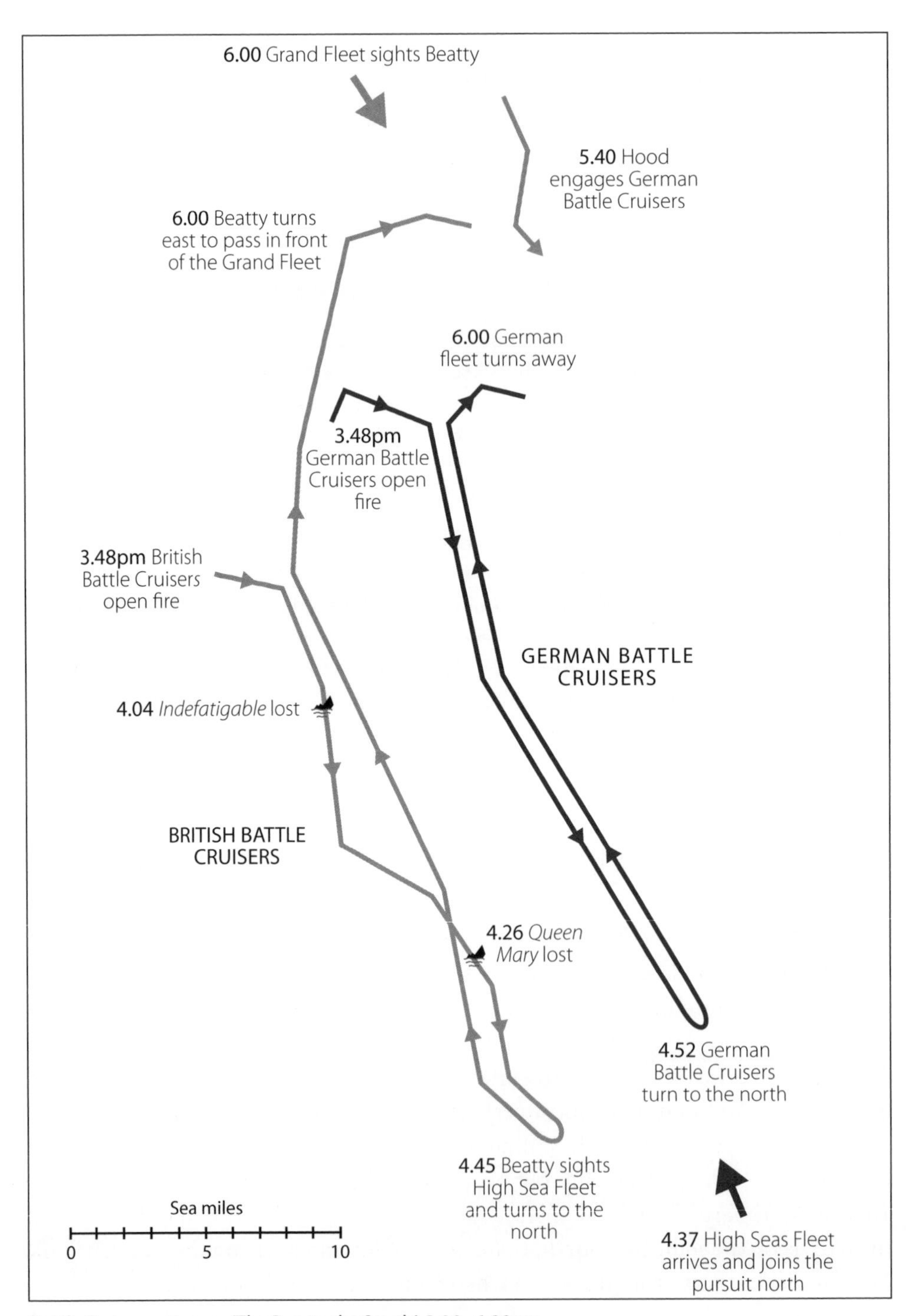

Battle Cruiser action, or 'The Run to the South', 3.30–6.00pm.

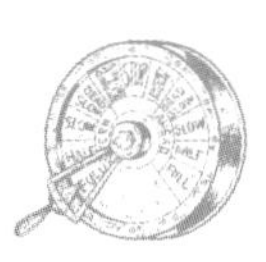

4.40 19½ kts
5.30 sighted barque bearing SE co NW
5.40 four smokes South (smoke screen 3rd LCS)
5.55 a/c as req for taking station on battle fleet

GENERAL PROGRESS OF THE BATTLE

So at 5.00pm Scheer with his main body of sixteen modern dreadnoughts and six less substantial pre-dreadnoughts, ten miles southward of the British, was now chasing the British to the north to support his second in command Hipper. This duel with the 5th Battle Squadron snapping and biting at the German van continued northward until 6.00pm. Both German and British ships sustained shell damage, though none was sunk in this period.

At 6.00pm, just as *Caroline* was taking station on the Grand Fleet, Beatty too met the leading ships of the British line. His depleted force was now toughened up with the addition of three more battle cruisers from the 3rd Battle Cruiser Squadron, detached from the Grand Fleet. The five doughty German battle cruisers turned away to the south-eastward. But at 6.31pm one of the detached battle cruisers from the Grand Fleet came to a catastrophic end. HMS *Invincible* was hit. Her magazine blew up. She sank, taking over 1000 souls with her. It is highly unlikely that any submarines from either side were present.

It will be recalled that Jellicoe, steaming down from the north (some miles to the rear of *Caroline)* with his combined force of twenty-four dreadnoughts, had rolled out in single line ahead formation. This was Jellicoe's great moment, the climax of a highly successful career. He remained calm. He had been informed the entire German Fleet was out. Victory was within his grasp that afternoon. Had not Winston Churchill stated that Jellicoe was the only man who could lose the war in an afternoon? So it was vital for Jellicoe to get the next two moves right. The first move was to convert his line ahead formation into 'cruising order', yet not too early and definitely, most certainly, not too late. It entailed forming into six divisions of four ships each, sailing in parallel. This essential preliminary move now saw *King George V* take up her position on the east flank as the first division. Then to starboard, *Orion* led the second division. *Iron Duke* as flagship led the third division, *Benbow* the fourth, *Colossus* the fifth and lastly on the western flank *Marlborough*. It was around 3.45pm when Jellicoe managed to get them sorted.

For the second move Jellicoe had to choose the right moment to deploy, to cast

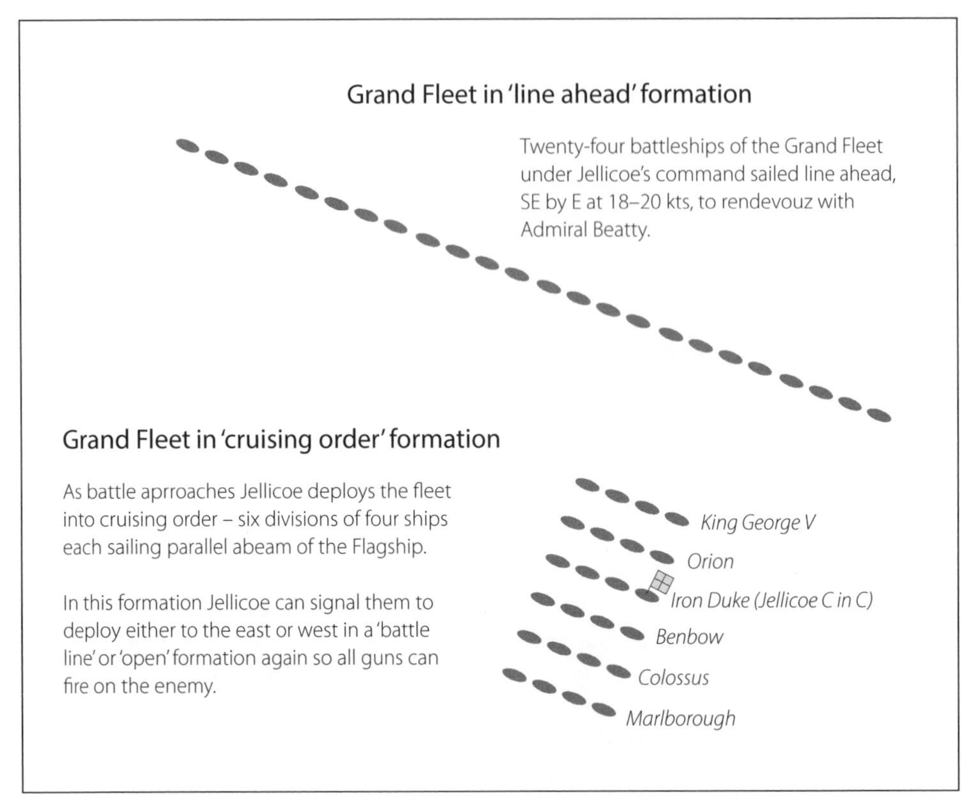

Diagram of the Grand Fleet in 'line ahead' and 'cruising order' formations prior to engaging the German High Seas Fleet.

his ships either to his west or to his east. In the mist and funnel and gun smoke of battle would he be able to divine the angle of the unsuspecting, on-coming German host to decide which way to deploy? If he timed this move right he could cross the enemy's 'T'. By this is meant: to deploy all ships into an open formation in which all guns can be brought to bear upon the enemy, thereby causing maximum damage to them. He had to get astride the enemy vanguard, to cross his 'T'.

Thus it was at 6.14pm that Jellicoe sighted the leading ships of the German fleet. He deployed from cruising formation into battle line to the east, one ship following in the path of the next ahead in a south east direction. From 6.00 to 6.30 pm the British battle fleet shaped itself into a curved line. Jellicoe had timed his deployment and its direction just right. Suddenly, the intimidating sight of such a mighty array of ships ranged against him as far as his eye could see came as a deep shock to Admiral Scheer.

Caroline's log again:

Wednesday 31 May 1916 [continued]

PM *6.07 enemy projectiles falling near ('overs' meant for BCF)*

6.15 one enemy light cruiser on fire

6.22 destroyer British hit by shell 'over' from BCF

6.31 BC Invincible *blew up (2 mile on starbd bow), bow and stern visible*

6.36 Battle Fleet opened fire

6.45 s/c WSW, 6.51 s/c SSE

6.55 16 pts to starbd [Caroline turns 180° to right]

7.06 co S 15° W

7.16 2nd Battle Sqdrn opened fire [astern of Caroline: King George V, Ajax, Centurion, Erin, Orion, Monarch, Conqueror, Thunderer*]*

7.13 fire observed in Kaiser class vessel

7.20 German smoke screen began [prelude to torpedo boat attack; Scheer decides to turn away under cover of his attacking torpedo boats]

7.22 torpedo attacks by German TBD

7.30 ship commenced firing at right hand enemy destroyers

7.34/35 torpedoes passed close to ship

7.47 turned to ram reported submarine, felt concussion under bottom

8.15 1st division 4th LCS and half 11th destroyer flotilla turned to attack enemy TBDs

8.17 firing commenced southward

8.20 Calliope *opened fire*

8.28 passed thro' quantity of wreckage, splintered wood on Carley life raft[4]

8.35 felt concussion under bottom

9.05 fired two torpedoes at small enemy battleship range about 7,600 yds

9.09 enemy's shell commenced to fall around ship. Full speed and stood out of range making smoke screen, 317 revs

9.30 took station ahead of Iron Duke, *co south 18 kts, 295 revs*

10.00 16 kts

10.37 battle fleet firing astern.

---

4 An easily launched, lightweight, rigid life raft used by many navies in both world wars.

GENERAL PROGRESS OF THE BATTLE

This episode, starting at 6.00pm is known as the Battle Fleet Action. The tempo of the battle increased to a frenetic pell-mell struggle, a period known as 'Windy Corner'. Beatty passed across the front of the Grand Fleet. The Grand Fleet deployed. HMS *Defence* blew up. HMS *Warrior* was disabled. At 6.17pm the British battle line opened fire. The cruiser *Wiesbaden* lay crippled and burning and at 6.20pm *Warspite*'s helm jammed. She found herself turning in a circle ever nearer to the German line.

Invincible exploded and went down at 6.31pm. At 6.54pm *Marlborough* was hit by a torpedo. Between 7.00pm and 7.14pm the remaining six British battle cruisers, some 5 miles ahead of the British line, succeeded in renewing contact with the enemy at an approximate range of 15,000 yards. At 7.10pm Scheer ordered his destroyers to launch an attack (with torpedoes) on the British battle fleet to cover his *Gefechtskehrtwendung,* his battle fleet's turn-away. As we see In *Caroline's* log there is mention of the German smoke screen at 7.20pm. Jellicoe decided not to follow the German turnaway in the belief that he would be drawn over mines and submarines.

Even though *Wiesbaden* was beyond rescue, at about 7.20pm a second German destroyer attack took place to cover a further German turnaway, five minutes later. At 7.45pm British battle cruisers lost contact with the enemy. About 8.20pm British light cruisers located the enemy again. At 8.20pm *Calliope* engaged the enemy. At 9.05 *Caroline* fired torpedoes at enemy ships. By 9.32pm the day's action was effectively over.

Albion Smith's diary reported *Caroline* going to action stations at 3.00pm. We resume his narrative of Wednesday 31 May 1916: "We were soon in the thick of it. It was a grand yet awful sight. The huge guns spouting fire and clouds of shrapnel bursting all around. The battle continued all around us with slight lulls in between. One destroyer had its funnel carried away. Then shells began to fall all around us. The enemy sent three torpedoes at the 'Carrie' but missed. We saw some of the enemy on fire and in sinking condition. Also saw the tragic end of one of the big ships and later steamed over the spot where she went down… [a reference to *Invincible*] It was an awful sight. After it got dark we came on some of the enemy battle ships and fired two torpedoes at them, one of which took effect. They fired two torpedoes at us but missed…"

At twilight on Wednesday 31 May 1916, it can be seen from *Caroline*'s log that an opportunity arose for taking the battle directly to the enemy. Commodore Le Mesurier's squadron, the flag ship *Calliope* and *Comus,* had already found themselves in a gun duel with more heavily armed battleships. *Calliope* was

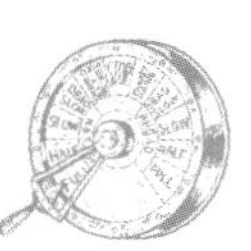

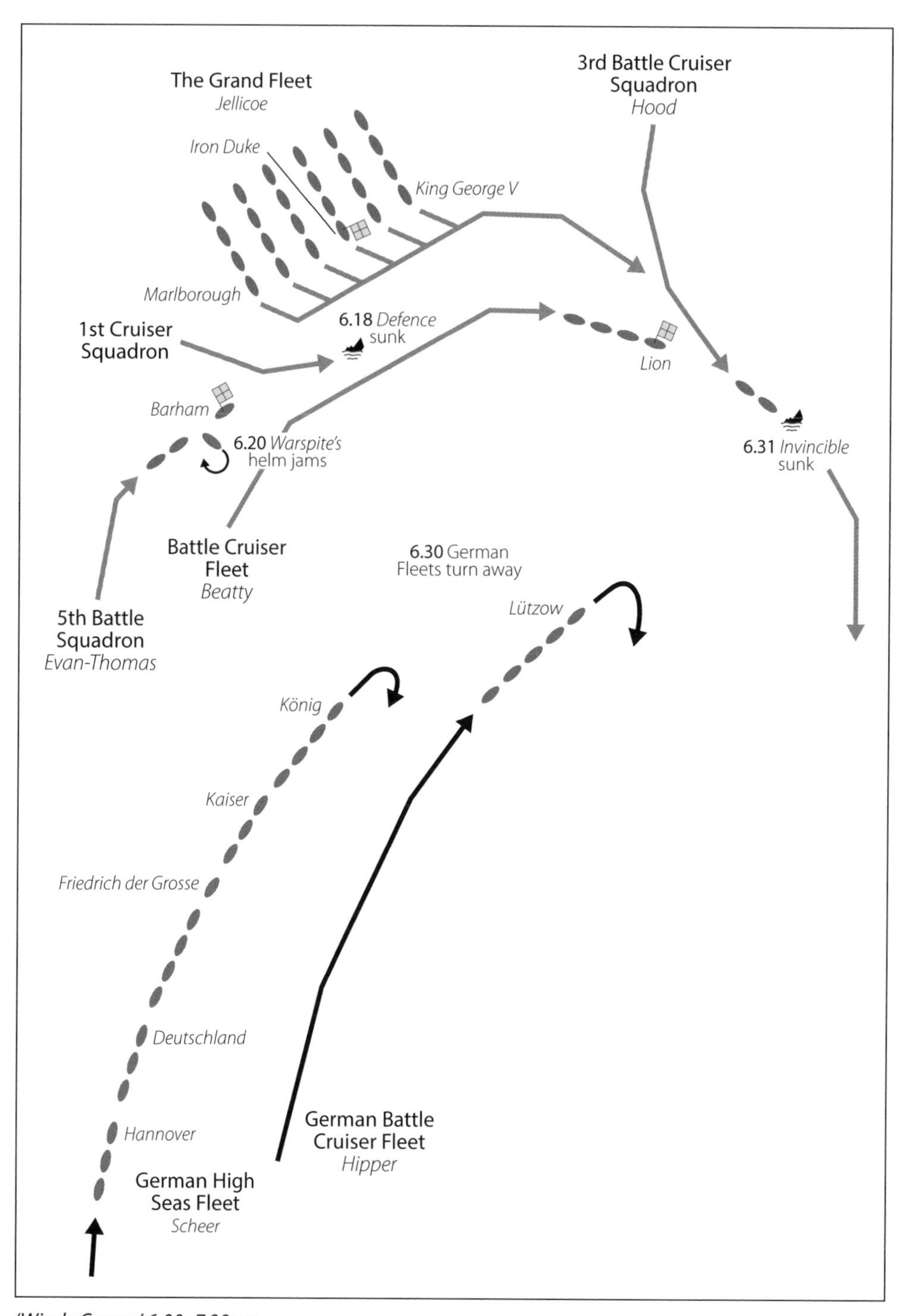

'Windy Corner' 6.00–7.00pm.

damaged but succeeded in firing a torpedo as she made her withdrawal. At 8.45pm *Caroline* and her sister *Royalist* sighted the shapes of three battleships against the darkening western sky. They hurried to identify the strange ships. Captain Crooke by 9.00pm was clear that they were German, supported by the belief that the distinctive shape of their upper deck cranes was typically German and not of the type used by the British.

Captain Crooke moved to the attack. But Admiral Jerram doubted the identity of the reported ships, supposing they were Beatty's battle cruisers. A message was flashed to abort the attack. Only when Crooke confirmed that he was certain did Admiral Jerram signal: 'If you are quite sure, attack.'

After this delay the attack went in at 9.05pm at a range of 7,600 yards and as the two ships altered course round to launch their torpedoes they received a hailstorm of shells from the battleships *Westfalen* and *Nassau*. Miraculously the two resourceful light cruisers escaped at high speed unscathed under cover of their smoke screen. It is said that one of the torpedoes made directly for the *Nassau* but, running too deep, passed under the ship. Admiral Jerram, still plagued by doubt, refrained from following up the enemy's withdrawal. As a result with nightfall all chance of a decisive result to this great North Sea battle evaporated.

In this account of *Caroline*'s particular experiences at Jutland it is not intended to look at the wide ranging controversy that was sparked by this mighty clash of arms between two rival sea powers. However, it is clear that in respect of *Caroline* and *Royalist*'s eleventh hour opportunity to turn the tables a great chance was missed. Vice-Admiral Sir Martyn Jerram's 1st Division of the British 2nd Battle Squadron, *King George V, Ajax, Centurion* and *Erin* failed to take the initiative and press home an overwhelming attack with their superior 13.5 inch guns against an enemy unable to match such fire power with 11 and 12 inch guns. It was a defining moment in that the issue might have been decided by British gunfire alone to the Commander-in-Chief, Sir John Jellicoe's, entire satisfaction.

To appreciate a more detailed first hand account of *Caroline*'s part in the battle we can draw on the second-in-command, Lieutenant Commander The Hon Drummond's stirring report on Jutland, May 31, as written out immediately on return from action:

> "3.30pm Exercised action and prepared for a scrap. About 5.40pm we sighted smoke on the starboard bow, and shortly afterwards heard sounds of firing in that direction; then the 2nd Light Cruiser Squadron and 1st Light Cruiser Squadron, with battle cruisers behind, came across our bows heavily engaged. We turned and followed, and being on the disengaged side of battle

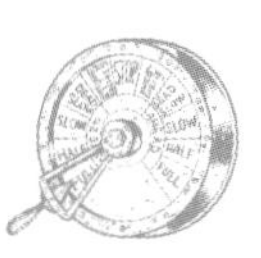

cruisers got all the German 'overs' among us, three or four shots falling just short and just over, and two destroyers astern of us got hit. We were all badly bunched, light cruisers and destroyers together, and it took some time to deploy and get clear, so it was lucky that more damage was not done to us. A little later the *Invincible* blew up ahead of us, and after this we deployed out and were on the engaged side of the battle fleet, but action near us broke off, although heavy firing could be heard and seen in the rear.

At this time one enemy cruiser appeared to be sinking by the stern, and after a few minutes she vanished completely. We also passed a German destroyer floating upside down. Shortly afterwards the fleets drew together again, and our battle fleet opened a heavy fire, *Iron Duke*'s squadron firing right over us. We next went ahead and moved out towards the enemy, to stop a torpedo attack which they were launching, and opened fire on a German destroyer. We had a very exciting few minutes dodging a torpedo, which came straight for us running on the surface, and then appeared to follow us round, missing our stern by inches. Another torpedo passed just ahead of us. About this time we rammed something, and as a submarine had been seen just previously right ahead of us, we hoped we had got her.

The destroyer we were firing on disappeared; several people say she went down, but it is hard to judge in the mist, although a lot of our shots appeared to be falling all round her. After this there was a lull, and we started edging over towards the enemy again until we made out a line of ships in the mist – we were stationed just ahead of *King George V*, leading the van battle squadron. They challenged us by flashlight, but made a wrong challenge, and when we saw this, and also were able to make out their cranes, which only Germans carry, we reported them as enemy, and at a range of about 5,200 yards fired both our torpedoes. *Royalist*, astern of us, also fired one torpedo. We then turned away at full speed and made smoke to cover ourselves. The ships we fired at were ships of the *Deutschland* class, we thought, and immediately we turned they opened fire on us, firing star shells once or twice. They straddled us at once and for the next eight minutes with 11-inch shell, but we were not hit although we had some close shaves, one shell passing between the wireless and upper deck, and spray from the splashes fell on the upper deck several times. A torpedo passed close to us at this time, and we twice bumped into something below surface.

After this we took up station for the night, rejoining our flagship, *Calliope*, which had taken her division away previously to make a torpedo attack

separate from us. We heard heavy firing astern of us about 10.00pm, and again about midnight, but we saw nothing more except for a large flare-up, a sheet of flame some way off. We were, of course, at action stations all the night and morning, sleeping at our stations.

At daylight there were no signs of the enemy, and though search was made all the morning they could not be found, and they must have got back to their ports. At 4.30pm on June 1st we packed up from action stations and returned to base. The conditions during the action were considered trying, as when we were first engaged and under fire we were too far off to be able to retaliate, and also were masked by other ships.

Except for firing at a destroyer and for making one torpedo attack, we did not get an opportunity of taking much hostile action. The behaviour of the ship's company was excellent; they were very cool, and cheered each time a hit was observed on the enemy. From observations on board, it appeared that just after the *Invincible* blew up, an enemy cruiser was heavily hit, and was sinking by the stern; otherwise, except for seeing a destroyer bottom up, and noticing another destroyer we engaged getting it fairly hot, and noticing some hits on enemy's battleships, not much could be seen of results. The smoke screen that we made with funnel smoke after our torpedo attack seemed very effective. But we were very lucky on this occasion, as the enemy got the range of *Royalist* and ourselves with their first salvo, and straddled us well, yet we were not hit.

Some special points that were noticed were that our guns seemed to be hitting, while German shells were all falling short; the distinctive black smoke when enemy shells exploded; the small splash enemy shells made compared to ours; and that the noise of the enemy shell bursting over the ship sounded exactly like a heavy gun firing."

In summary, as daylight faded, both *Caroline* and *Royalist* had discharged their duty as light cruisers in the face of the enemy: to scout ahead, correctly identify the enemy forces in their vicinity and report back. Precious moments were spent convincing senior officers up the chain of command of the true identity of the ships within their range. Shackled by indecision and doubt Vice Admiral Sir Martyn Jerram refrained from following up the light cruisers' bold probing initiative by pressing home an attack with his powerful division. A Nelson moment squandered, let us note Sir Julian S Corbett's comments in his official naval history of the Great War:

"The *Caroline* and *Royalist* had just seen what they took to be the German pre-dreadnought squadron, and the senior officer Captain HR Crooke made the signal to attack with torpedo. Admiral Jerram who still could not make out where Admiral Beatty was and was expecting to sight him any moment at once signalled: 'Negative the attack. Those are our battle cruisers'. Captain Crooke, however, could see more clearly and having no doubt as to what he had sighted took upon himself the responsibility of ignoring the order and proceeded with the attack. In spite of the storm of fire that met them the *Caroline* fired two torpedoes and the *Royalist* one at 8000 yards. Then smothered with shell they made off, though both were straddled again and again…"

18

Black Night Over Water

During the action of 31 May, especially at the time of 'Windy Corner,' visibility had deteriorated toward the early evening. This was due in part to mist and haze. But conditions were also made worse by smoke, belching funnel smoke, protective smoke screens and the discharge of hundreds of guns. An otherwise calm sea was churned up by a myriad of ships' wakes criss-crossing each other. By 9.00pm the British battle fleet had taken up night dispositions. The Commander-in-Chief, Admiral Jellicoe, to whom a continuation of the battle at night was unacceptable, decided to shape a southerly course with a view to re-engaging his enemy at daybreak. British destroyers with light cruisers were sent back through the Grand Fleet to cover the British battle line's rear five miles astern, and to block an eastward quest home by Scheer.

Caroline was fortunate in yet again avoiding damage or outright destruction by dint of her forging southward with the 4th Light Cruiser Squadron ahead of the Grand Fleet. So it was that a confused, highly dangerous close range conflict at night behind the Grand Fleet marked the final phase of the Battle of Jutland. It saw the German battle fleet neatly escape eastwards toward the position of the Horns Reef light vessel, and from thence a safe deliverance home. Admiral Scheer achieved this conjuring trick by slowing his pace southward to sixteen knots, while Jellicoe sped onwards at seventeen knots. For lack of speed Scheer could not cross ahead of the British battle fleet. He chose to drive south-eastward toward the Horns Reef. It enabled Scheer to steal across the wake of the Grand Fleet. As seen earlier the final log entries for 31 May found the 4th Light Cruiser Squadron, including *Caroline*, in station ahead of the flagship *Iron Duke* on a course south at 18 knots, 295 revs. At 10.37pm the Battle Fleet heard firing astern.

Thursday 1 June 1916

AM *2.30 a/c 16 pts to stbd. Formed S.L. abreast to Port, 5 cables apart*
2.50 zigzag 1 pt

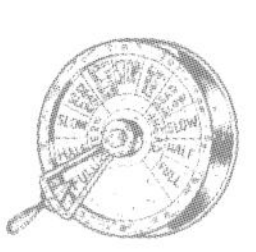

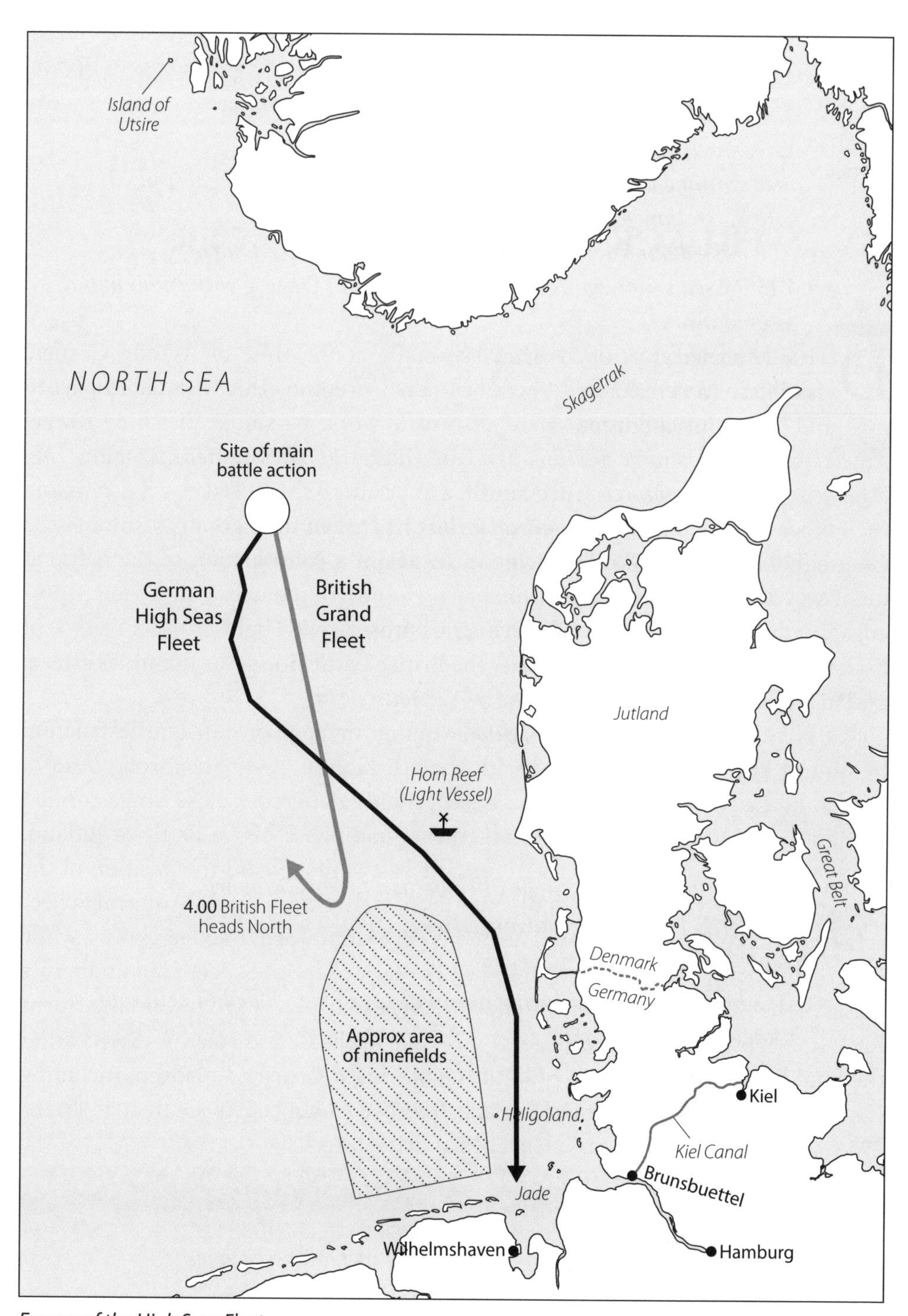

Escape of the High Seas Fleet.

3.30 17 kts
3.53 turned together as req
4.10 a/c NNW 17 kts
4.40 North 18 kts, zigzag 1 pt. each way
5.20 stationed ahead of B fleet 3 miles co. N 17½ kts, zigzag 1 pt.
6.10 SE stationed abeam 17 kts
7.00 passed 3 masted schooner steering NE Europa, *Dutch*
7.15 Passed small Danish fishing boat Toye of Esbjerg *with small motor boat away*
7.20 turned 16 pts increased 20 kts
7.45 co North 20 kts
7.53 repassed Europa
8.05 passed large quantity of oil on water – 3 bodies in lifebelts, one lifebuoy marked 'Turbulent'
9.00 a/c as req with squadron for taking station
10.18 co N by W 17½ kts, zigzag 1½ pts
10.32 hands to action stations
11.45 a/c NW reformed abeam of Comus, *zigzag 1 pt.*

PM *2.05 a/c as req 2.25 res co N by W, 15 kts zigzag*
3.35 a/c as req for taking station
3.45 s/c North 15 kts
4.50 a/c N 20° W
5.50 a/c North
6.00 formed S.L. AB [single line abreast] to Port 5c apart
6.10 s/c N by W 16 kts, advance 16 kts
6.20 zigzag 1pt
7.12 halfmast colours during burial at sea of men killed in action in Malaya *and* Barham
8.00pm 2 miles ahead of Battle Fleet
9.45 in station astern of Royalist *N 59° W 19 kts*

Friday 2 June 1916

AM *0.32 expended Lyddite rounds in four 4 inch and 6 inch guns*
0.50 Y4 berth, recd 423 tons oil and ammunition to complete

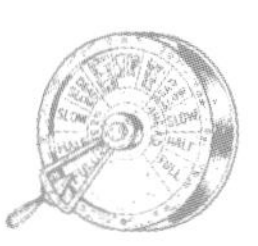

Thursday 1 June 1916 Albion Smith writes: "...very little sleep that night, lay down in our clothes, out at action stations next morning (Thurs) at 2.30 (firing had been going on in our rear all night). We were called to action stations several times during the day. It was a trying time for all of us, no chance of a wash, very little to eat and still less sleep. At 7.15 we all stood at salute with our flags at halfmast while other ships buried their dead. Came on to pour with rain later in the evening. Ship rolled terribly in night and worse in the morning. In fact it seemed as tho' she would capsize at any moment. We got near Flow about 10pm waited outside until all the big ships had passed thro and we reached anchorage about 12.30. Then the hospital ships came alongside the damaged battle ships and took off the dead and wounded. Very sad it all seemed and rather unreal."

As we have seen, the final episode of the Battle of Jutland took place after dark – in the pitch black of a moonless night. The German fleet managed to find its way home without re-encountering its enemy, effectively by crossing the wake of the Grand Fleet and heading eastwards.

Unlike peacetime no lights could be shown to enable ships to steer clear of each other and avoid collisions. Communication between one ship and another, whether friend or foe, was fraught with risk. Visual signalling by semaphore and flag hoists were an established daylight method of keeping in touch; absolutely out of the question now after 10.00pm. Nor was radio telephony an option – a development that lay far in the future. The matter of life or death now lay either in the astute or the unwise use of the searchlight.

An agreed but secret challenge, like a time-limited valid password could be used to identify a strange ship as a friend upon a correct reply or as a foe upon an incorrect response. But what if the enemy already by some freak circumstance possessed the correct British challenge and reply or even merely a part of it? Such a thing would be unthinkable. But at 9.30pm this very circumstance was allowed to take place. HMS *Lion* flashed a signal to HMS *Princess Royal* asking for the valid challenge and reply, as *Lion*'s had been lost. *Princess Royal* responded by giving the information needed by flashlight. This was most unwise because just two miles distant invisible German light cruisers were able to make out the gist of the letters signalled by flashlight. The enemy was to use the information to lethal effect on this blackest of nights. A flashed challenge could of course, in an instant pinpoint one's position. One only has to imagine how any living body feels when suddenly dazzled at night by the stark glare of headlights. The effect can be stunning, disorientating and terrifying.

While *Caroline* and the 4th Light Cruiser Squadron forged ahead of the Grand Fleet, now in night cruising order on its southward course, the 4th Destroyer Flotilla

was directed back to take station five miles astern of the Grand Fleet to shield its rear. Some distance ahead of the 4th Flotilla was the 5th Battle Squadron, following the Grand Fleet southward. Unknown to them, much nearer to the 4th Flotilla and on its starboard side, was the German battle line questing unwaveringly eastward as per Scheer's express directive: "Kurs SSO ¾ O, 16 Seemilen Fahrt. Durchhalten" (Course south south-east three quarters of a point east, 16 knots. To be maintained.)

The 4th Destroyer Flotilla was lead by the destroyer leader HMS *Tipperary*, 2,137 tons. The Captain in overall command of the flotilla formed his ships in single line ahead; *Tipperary* leading the first half flotilla: *Spitfire*, *Sparrowhawk*, *Garland*, *Contest* followed by *Broke*, *Achates*, *Ambuscade*, *Ardent*, *Fortune*, *Porpoise* and *Unity*.

As chance would have it the Flotilla Captain was CJ Wintour, the officer with whom an exchange had taken place in January 1915. Crooke had left *Caroline* to spend that month in *Swift* and Wintour had assumed command of *Caroline*.

It was now just after 11.00pm when Wintour's half flotilla, on a knife edge of alertness, with all guns and torpedo tubes trained, made out ships to starboard. Were they British or were they foe? The silhouettes were large and could not be destroyers. The strange ships were closing. Should Captain Wintour first challenge them before attacking to establish their identity? How could he confidently take a chance and assume these ships were enemy? The unfortunate Captain Wintour could not have known the recognition signal had already been compromised. So he did what his conscience dictated. *Tipperary* made the challenge. The response was to be bathed at once in the stark glare of search light beams amid a crippling storm of fire. *Tipperary's* bridge was destroyed and all in it slain. Soon Wintour's ship was reduced to a wreck. In the general confusion *Spitfire* tried to rescue survivors. While in a state of temporary blindness due to the flash of guns and the glare of searchlights, *Spitfire*'s commander suddenly made out the awesome sight of an enemy cruiser bearing down upon him. At such close quarters his diminutive command was no match for a cruiser. Commander Trelawney swung his ship's wheel to starboard to avoid an outright collision so that the port side of the cruiser served a glancing blow to the port side of *Spitfire*. The two unequal sized ships scraped past each other. The blast from the enemy's guns was sufficient alone to devastate *Spitfire's* bridge. She later reached home port with a sizeable length of plating adorning her side.

Other equally ferocious close encounters took place that black night in which the side that challenged first lost the advantage of getting in the first salvo.

From *Caroline*'s log we see as she made her way home how she passed floating debris, poignant evidence of a macabre dance of death at sea, the battle that

took place off Jutland Bank. Amongst that debris, two solitary life buoys marked *Turbulent*, a British destroyer that had found herself in the path of the German battle line at 1.00am on 1 June. Her total destruction was sealed by the German battleship *Westfalen,* firing into her at point-blank range. The tiny ship was blown to pieces.

Caroline had delayed her return to her usual anchorage until the early hours of 2 June, the ships with the wounded and the dead having priority and being attended to by hospital ships. Albion Smith felt "very sad it all seemed and rather unreal". However, he and his messmates were allowed no time to relax from the adventures of the past forty-eight hours, as the 4th Light Cruiser Squadron's workhorse was summoned again to sea. The signal was flying: 'Light Cruisers To Sea'.

Albion Smith writes: "Dull morning Saturday – the light cruiser *Royalist* and ourselves left Scapa at 8.00am to go on patrol duty, also to look for a reported German armed raider. We had a submarine scare during the day. We sent away a boat to board two Norwegian ships, one named the *Solborg* out of Buenos Aires. The ship rolled only slightly on our way north. By 10.00pm we were off the Faroe Isles. Sunday dawned a fairly fine morning after a rainy night. We held up and boarded a Swedish whaler, though nothing to report. Later passed a fleet of Norwegian fishing smacks… we passed one of our armed liners – two funnels and three masts".

All Sunday night *Caroline* endured vile weather and Smith continues: "Rolled badly all night and still worse next morning, raining all the time. Overhauled a steamer about 9.00am but did not board her. Later on passed close to Dutch steamer *Rangoon*, 4,360 tons and the steamer *Iberia*. Weather bad all day waves breaking over the ship and rolling badly. All of us on tinned food, fresh meat and spuds having run out on Saturday. Very rough all night, ship rolled terribly. No one allowed on the upper deck."

On Tuesday 6 June 1916, Albion Smith writes "Wet miserable morning, we reached Scapa at 12.30 and got a big mail on board and also long looked for news of naval fight…'

To those men of the Fleet far from home in remote Scapa and parted from their families for endless weeks and months at a time, we can imagine their delirious rejoicing at the sight of the mail boat drawing alongside and the consequent uplift of morale as letters and parcels from home were distributed. Albion received mail regularly from his school friends and colleagues in the Merchant Navy; not always good news but sad as well after reports of friends killed in France. His relationship with Chrissie, whom he had last seen in Brighton, prospered as on 12 June he received a letter from her with a rose enclosed. What better symbol could one have to brighten a "dull morning, cold wind, and big sea running."

In forming a view of HMS *Caroline*'s place in the Fleet and her role in the recent battle we are helped by the testimony of Lieutenant Commander Drummond, as set out in these pages, and that of the ship's war log, crowded with minute by minute detail still in a legible state today. We have seen also from the judgement of Sir Julian S Corbett, the official naval historian of the Great War, how Captain Crooke boldly took upon himself the decision to seize the moment to attack the *Westfalen* and *Nassau*, which meant in effect ignoring his superior's implied instructions.

Let our diarist Albion Percy Smith have the last word. A few days after the action he penned a fuller description of it for his family. It is clear that the experience of witnessing the battle unfolding in real time from the vantage point of *Caroline*'s bridge made a deep impression on him, inspiring him to expand on his original reflections:

> "... the sea was dead calm with only a slight wind and occasional gleams of sunshine, the temperature being slightly chilly with a haze hanging round the horizon. We knew Norway was somewhere near. About 4.00pm we heard the sound of distant guns. Imagine our excitement and joy when we knew that after months of dreary watching and waiting we were soon to have a go… Once in range the sights and sounds which met our view were too varied to be described. It was a grand but awful sight… we were ahead of most of our fleet this being due to our speed and purpose for which our ship was built. Then the huge guns of our battleships opened fire… Added to this was the din of the fight already taking place. Guns were hurling 15-inch shells into the opposing fleet with roars and flashes as if scores of thunderstorms had met and got angry. The sea became churned into waves and foam, this being caused by the speed and movement of scores of ships of all sizes…"

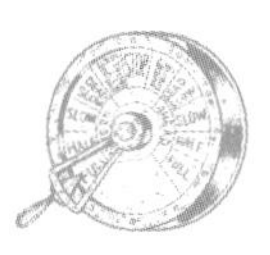

EPILOGUE

In the course of her service in the months preceding Jutland and in its aftermath, HMS *Caroline* had steadily earned for herself a reliable reputation as a 'can do' warship. Her ship's company were rightly proud of her. Shortly after Jutland Captain Crooke, amongst other senior officers, received an invitation to dine with the Commander-in-Chief in his flagship HMS *Iron Duke* – a 'wash up' following the excitement of the great battle.

The 19 August 1916 saw the captain married at St Stephen's church Kensington. Albion Smith was invited to the reception. On 15 September 1916 Captain Crooke was 'commended for services in Jutland battle' by Commodore CE Le Mesurier, and on 13 October 1916 *Caroline* added laurels to her reputation. It was the occasion of a big boat race for the whole fleet. Captain Crooke won the cup and gave a supper to his boat's crew for helping him.

On the 3 February 1917 *Caroline* sailed to the Clyde for a refit at Fairfield's Yard, Govan. It changed the ship's profile. In her original plans she was armed with eight single 4 inch guns – two each side of the bridge and two each side of the waist – two single turrets aft of the centre line, one single 3-pounder in 'Q' position (amidships) and two twin 21 inch torpedo tubes. Her upgrade saw the four forward guns removed and replaced with one 6 inch turret on the centre line and two 3 inch high angle guns each side of the bridge to counter the new menace of the aeroplane. The 3-pounder in 'Q' position was replaced with a further 6 inch gun. The single original pole mast was changed to a tripod, supporting a 'fighting top'.

On 27 March 1917 Captain Crooke left the ship to take up an appointment at HMS *Excellent,* Whale Island. Albion Smith stayed to pack and clean up. On 2 April Smith settled in at Whale Island resuming his duties as captain's steward. By way of welcome, all Captain Crooke offered him was a raised eyebrow and the wry comment: "So you've arrived, Smith."

On 11 June 1919 Captain Crooke was awarded Companion of the Order of the Bath (CB) for valuable services with reference to Director of Naval Ordnance. Subsequently he commanded the battleships *Benbow, Empress of India* and *Marlborough*, ending his career in the rank of Rear Admiral in 1942.

Admiral de Robeck said of him: "...he should make a good admiral afloat or in a shore appt. His ship is well handled. Ability above average."

The remainder of the Great War saw *Caroline* patrolling the North Sea. After

the end of hostilities *Caroline* was re-commissioned for service in the East Indies Station where she served from 1919 to 1922, after which she was paid off into reserve. Next, Sir James Craig, in the guise of a 'white knight', rescued *Caroline* to give her a promising future from 1924 as the headquarters ship for the Ulster Division of the Royal Naval Volunteer Reserve (RNVR) in Belfast. Meanwhile one by one from 1921 to 1927 seven of *Caroline*'s running mates were sold, 'surplus to requirements'.

During the inter-war years the Ulster Division of the RNVR steadily built its numbers; monthly orders for 1933 recorded a complement just short of 500 officers and ratings. Of particular value to the new Ulster Division were de-mobbed Royal Navy personnel with active sea experience gained during the late war; people with specialist skills in the sphere of, for example electrical, engineering and seamanship ratings. One founder member with a set of skills invaluable to the freshly created Ulster Division was Leading Telegraphist White. He had served in the Navy as a Boy Signalman from 1917/18.

He was mobilised in 1938 during the Munich crisis and again in 1939. De-mobbed in 1947 he rejoined the Ulster Division and retired in the rank of Commander in 1955. His son, Brian White, joined the Division as a rating in 1949. He was promoted to midshipman in 1951 and commissioned sub lieutenant in 1952. Thus it was that for a time both father and son had the distinction of being members of the same wardroom – *Caroline*'s wardroom.

In 1937 during the Coronation tour Their Majesties King George VI and Queen Elizabeth visited the ship and inspected her ship's company. They were received on board by the commanding officer, Captain The Earl of Kilmorey and Commander Richard Pim, the executive officer (second in command). Commander Pim, later Captain Pim, in the 1939–1945 war held a unique and singularly important post as senior officer in charge of Winston Churchill's map room. Today the valuable work this Ulsterman performed for the war effort is commemorated in the Cabinet War Rooms, Whitehall. They are open to the public.

During the course of the Second World War, HMS *Caroline* was moved from the Musgrave Channel in the Port of Belfast to a berth in the Milewater Basin and managed to emerge totally unscathed after two major raids by the Luftwaffe. She was used as a trawler base, providing facilities for storing and fuelling ships returning from convoy work in the Western Approaches. Meanwhile two of her commissioned officers, Sub Lieutenant Ivan Ewart and Lieutenant W Stevens distinguished themselves in the fighting war at sea. The former commanded a motor gunboat in the Narrow Seas (English Channel and southern North Sea) and the latter a similar vessel on the occasion of the daring, surprise attack on St Nazaire. Both were captured and were privileged to spend time in the German

maximum security prison, Colditz. 'Billie' Stevens escaped and managed to make a home run.

After the war HMS *Caroline* resumed her pre-war role as a base for the Ulster Division RNVR. The Division acquired a modern, 360 ton minesweeper, *Kilmorey*, which was used for sea training. The year 1964 saw *Caroline* celebrate 50 years in commission. The Duke of Edinburgh and a select few elderly members of her original ship's company, including Albion Smith, were honoured guests on that day. What a heart-warming and wonderful experience it must have been for those veterans to find themselves once again aboard the very ship in which they had served their first commission. However, for those actually serving in the ship fifty years later in 1964, the author included, sentiment and pride in their ship was intermingled with grave doubt about the future. The author remembers that there was much speculation as to how long the ship could last. No one, not even the most positive of optimists imagined that *Caroline* could still count on another 47 years 'in commission' to the year 2011.

After 2011 it became clear that goodwill on the part of the people of Ulster toward HMS *Caroline* would not be enough alone to save the ship in time for the 2016 centenary of the battle of Jutland. However *Caroline* possessed a unique quality. Indeed of all the two hundred and fifty British and German warships that took part in the historic Battle of Jutland, HMS *Caroline* was now the sole survivor. *Caroline* held and will continue to hold the title-deed as direct witness to that battle, just as HMS *Victory* can claim the same accolade in her link with Trafalgar. By dint of still being afloat today this light cruiser, representative of her class, presents us with the last remaining historic link with that battle.

None the less certain expertise in the management of historic ship conservation would be needed as well as substantial finance. And how could a suitable, exciting interpretation of this historic light cruiser emerge for the benefit of the enquiring public? To the great delight of all, a new chivalrous knight was not long in appearing on the scene in the shape of the National Museum of the Royal Navy in partnership with the Department of Enterprise Trade and Investment. With wide experience of such issues, it is reassuring to know that this historic body will now pilot HMS *Caroline* to her next waypoint in her long life odyssey.

And so, the year 2016, one hundred years after Jutland, saw HMS *Caroline*'s ever growing glorious reputation rise to a new height as she received the award of the five-star visitor attraction from the Northern Ireland Tourism Board.

ACKNOWLEDGEMENTS

I OWE A life-long deep interest in HMS *Caroline* to my late father Richard Sydney Allison. During the Second World War he reached the highest rank that a medical officer in the Royal Naval Volunteer Reserve could aspire to, that of Surgeon Captain. In 1974, together with a committee of officers comprising Arthur Orr, Brian White, David Potter, Robert Shanks and Harold Shepperd serving in *Caroline* post war, my father completed a comprehensive history of the ship and the parallel development of the Ulster Division of the Royal Naval Volunteer Reserve, later Royal Naval Reserve. This work has been a valued point of reference when writing this particular account of *Caroline* and her doings in the years 1914–1917. I acknowledge with thanks the permission of the publisher The Blackstaff Press to reproduce material from the book.

I am indebted to William Meredith of Wirral Archives Service, Birkenhead and British Aerospace Systems for permission to reproduce illustrations from Cammell Laird's launch particulars of *Caroline* and other data relating to her construction. I wish to thank the staff at the National Archives, Kew, London, for their ready help in making the ship's war log available to me for my research. My grateful thanks are also due to Commander Roger Paine, RN Retired, who read parts of my manuscript and gave helpful guidance. To Jose Loosemore, daughter of Albion Percy Smith (Captain Crooke's personal steward) my special thanks for permission to use and quote freely from Smith's diaries, whose observations thereby bring life and lustre to the otherwise rather neutral entries in the ship's log. I am also grateful to Peter Bleakley for putting me in touch with Josie Loosemore and for his interest in the project. Also Derek Dray of the Chatham Dockyard Historical Society and Heather Johnston, Nick Hewitt and Andrew Baines of the National Museum of the Royal Navy, Portsmouth, who deserve my thanks.

Others who have been a source of close support, information and encouragement are my good friends Dr Graeme Kennedy, Roy Thompson, Allan Marshall and Dr Margaret Millman for her critical reading of the manuscript. My most constant and enduring guide throughout this project has been my partner Deirdre Howard-Williams who has patiently given of her time, advice and expertise at every stage in the preparation of my book. In addition Malcolm Johnston, commissioning editor at Colourpoint Books deserves my very best thanks for his steadfast enthusiasm and belief in the value of *Caroline*'s story from the start.

I have endeavoured to acknowledge all sources of help and information to the best of my ability. If I have unwittingly omitted anyone I do apologise and would be happy to make this good in any further editions.

APPENDIX 1

Yard No. 803 (HMS *Caroline*) Cammell Laird, specification

Building drawings received from Admiralty	21 August 1913
Admiralty order No: CP48997/13/x 14404	4 October1913
Completion date estimate:	21 May 1915
Delivery:	At the nearest open water to the port of construction
Penalty for delay in delivery:	£40 per wkg day
Contract returned to Admiralty:	30 January 1914

Trials:
Basin trial, preliminary trial, 30 hour trial at 4/5HP, 8 hours full power trial, stopping starting, astern & auxiliary machinery trial, 8 hours acceptance trial after opening up.

Estimate of cost:	£336,000
Keel plates laid down:	28 January 1914
Launched:	21 September1914
Delivered:	17 December 1914
Length overall:	446 feet
Length, between perpendiculars:	420 feet
Beam (width):	41½ feet
Draught:	13–16 feet
Displacement:	3,750 tons
Full load displacement:	3,800 tons

Armament

Forecastle:	4× 4 inch quick fire guns
Waist:	4× 4 inch quick fire guns Twin 21 inch torpedo tube mountings each side, amidships
Aft:	2× 6 inch guns, super firing
After engine-room casing:	3lb high angle gun

Protection

Side, amidships	3 inches
Bows:	2½–1½ inches
Stern:	2½–2 inches
Upper deck amidships, rudder head:	1 inch

Machinery

Geared Parsons turbines:	4 shafts, 30,000SSHP
Yarrow boilers, water tube:	8
Fuel capacity:	917 tons
Diameter propellers:	88 inches
Pitch of propellers:	78 inches

Surface:

Combustion space:	4,812 cu. Ft.

Sources:

Book No 2 Birkenhead, "List of Vessels built by Cammell Laird at Cammell Laird" "Launching Particulars", pages 45, 46. These records are now held by Wirral Archives Service, to whom I am grateful for access and for their permission to publish.

Author's note:

It is of note that the launch of Yard No. 813 on No. 3 slipway one year later (HMS *Constance*) was described as "quite satisfactory", while that of Yard No. 803 was "very satisfactory".

Boilers:

The 'Scotch' boiler was once used by naval vessels. It was in common use for railway locomotives and in the merchant marine. This type of boiler has a number of fire tubes through which pass hot gases from the furnace on their way to the funnel. The water lies around the tubes and steam is got up quickly.

In the water-tube boiler the water is in the tubes themselves, the flames playing directly around them. By this method steam is raised more readily.

By 1904 the Admiralty concluded after investigations that the water-tube boiler was best for naval purposes, placing orders with Yarrow and Babcock & Wilcox.

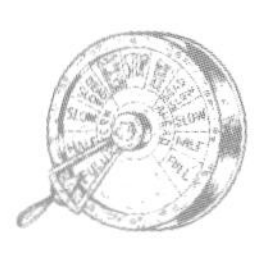

APPENDIX 2

Extract of Results of Four Hour Power Trial on River Clyde, Measured Mile

Yard No.803
14th December 1914

Draught, mean 13' 11½ inches, Final coat of paint applied 5th December 1914, High Water 14th December 9.00am, Barometer, 28.7" Weirs, Displacement on leaving anchorage: 3,955 tons. Three series "A", "B", "C":-

	"A"	"B"	"C"
Start:	9.14am	10.54am	1.08pm
Finish:	9.31am	12.15pm	1.23pm
Speed of Screw:	31.55	33.85	35.8
Speed of Vessel:	26.278	27.307	28.30
Horse Power:	24,615	31,402	36,083
Revs	491.88	528.48	558.38

Boilers	Air Pressure	No. Nozzles open	No. Burners in Use	Vacuum
Forward Boiler Room	1.87"	15		26.75
Aft Boiler Room	1.94"	13	54	27.15

Run "A" Prop. Revolutions, Pressure of oil 138, Temperature 193
Outer Stbd: 485, Inner Stbd: 497.5, Outer port: 485, Inner port: 500

Loss of water per 1,000 HP per 24 hrs	2.34 tons
Loss of lub oil " "	240 galls
Mean temp. Of turbine bearings:	135.5
Pressure of lub oil at pump:	35.8F 21A
Pressure of lub oil at bearings:	4.15F 1.31A

"machinery worked satisfactorily throughout trial"

Source: Wirral Archives Service

APPENDIX 3

The British Fleet at the Battle of Jutland

The Grand Fleet

Commander-in-chief: Adm Sir John Jellicoe (HMS *Iron Duke*)

Battleships

2nd Battle Squadron: VAdm Thomas Jerram
1st Division: HMS *King George V* (flagship), HMS *Ajax*, HMS *Centurion*, HMS *Erin*
2nd Division: HMS *Orion* (flagship), HMS *Monarch*, HMS *Conqueror*, HMS *Thunderer*

4th Battle Squadron: VAdm Sir Frederick Sturdee
3rd Division: HMS *Iron Duke* (Fleet Flagship), HMS *Royal Oak*, HMS *Superb*, HMS *Canada*
4th Division: HMS *Benbow* (flagship), HMS *Bellerophon*, HMS *Temeraire*, HMS *Vanguard*

1st Battle Squadron: VAdm Sir Cecil Burney
5th Division: HMS *Colossus* (flagship), HMS *Collingwood*, HMS *St. Vincent*, HMS *Neptune*
6th Division: HMS *Marlborough* (flagship), HMS *Revenge*, HMS *Hercules*, HMS *Agincourt*

Accompanying Cruisers

1st Cruiser Squadron: RAdm Sir Robert Arbuthnot
HMS *Defence* (flagship), HMS *Warrior*, HMS *Duke of Edinburgh*, HMS *Black Prince*
2nd Cruiser Squadron: RAdm Herbert Leopold Heath
HMS *Minotaur* (flagship), HMS *Hampshire*, HMS *Shannon*, HMS *Cochrane*
4th Light Cruiser Squadron: Cdre Charles Edward Le Mesurier
HMS *Calliope*, HMS *Constance*, HMS *Comus*, HMS *Caroline*, HMS *Royalist*

Accompanying Destroyers

4th Destroyer Flotilla: Capt Charles Wintour
HMS *Tipperary* (flotilla leader), HMS *Spitfire*, HMS *Sparrowhawk*, HMS *Garland*, HMS *Contest*, HMS *Owl*, HMS *Hardy*, HMS *Mischief*, HMS *Midge*, HMS *Broke*, HMS *Porpoise*, HMS *Unity*, HMS *Achates*, HMS *Ambuscade*, HMS *Ardent*, HMS *Fortune*
11th Destroyer Flotilla: Cdre Hawksley
HMS *Castor* (flotilla leader), HMS *Ossory*, HMS *Martial*, HMS *Magic*, HMS *Minion*, HMS *Mystic*, HMS *Mons*, HMS *Mandate*, HMS *Michael*, HMS *Kempenfelt*, HMS *Marne*, HMS *Milbrook*, HMS *Manners*, HMS *Moon*, HMS *Mounsey*, HMS *Morning Star*
12th Destroyer Flotilla: Capt Anselan John Stirling
HMS *Faulknor* (flotilla leader), HMS *Obedient*, HMS *Mindful*, HMS *Marvel*, HMS *Onslaught*, HMS *Maenad*, HMS *Narwhal*, HMS *Nessus*, HMS *Noble*, HMS *Marksman*, HMS *Opal*, HMS *Nonsuch*, HMS *Menace*, HMS *Munster*, HMS *Mary Rose*

3rd Battle Cruiser Squadron: RAdm Horace Hood
HMS *Invincible* (flagship), HMS *Inflexible,* HMS *Indomitable*

Accompanying Cruisers
HMS *Canterbury*, HMS *Chester*

Accompanying Destroyers
HMS *Shark*, HMS *Ophelia*, HMS *Christopher*, HMS *Acasta*

Battle Cruiser Fleet

VAdm Sir David Beatty

Battle Cruisers
HMS *Lion* (flagship), HMS *Princess Royal*, HMS *Queen Mary*, HMS *Tiger*, HMS *New Zealand*, HMS *Indefatigable*

Accompanying Light Cruisers:
1st Light Cruiser Squadron: Cdre Edwyn Alexander-Sinclair
HMS *Galatea*, HMS *Phaeton*, HMS *Inconstant*, HMS *Cordelia*
2nd Light Cruiser Squadron: Cdre William Goodenough
HMS *Southampton*, HMS *Birmingham*, HMS *Nottingham*, HMS *Dublin*
3rd Light Cruiser Squadron: RAdm Trevylyan Napier
HMS *Falmouth*, HMS *Yarmouth*, HMS *Birkenhead*, HMS *Gloucester*
Attached: HMS *Engadine* (seaplane tender with 4 seaplanes)

Accompanying Destroyers:
13th Destroyer Flotilla: Capt James Farie
HMS *Champion* (flotilla leader), HMS *Obdurate*, HMS *Nerissa*, HMS *Termagant*, HMS *Moresby*, HMS *Nestor*, HMS *Nomad*, HMS *Nicator*, HMS *Onslow*, HMS *Narborough*, HMS *Pelican*, HMS *Petard*, HMS *Turbulent*
9th Destroyer Flotilla: Cdr Malcolm Goldsmith
HMS *Lydiard*, HMS *Liberty*, HMS *Landrail*, HMS *Moorsom*, HMS *Laurel*, HMS *Morris*

5th Battle Squadron: RAdm Hugh Evan-Thomas
Battleships: HMS *Barham* (flagship), HMS *Valiant*, HMS *Warspite*, HMS *Malaya*

Accompanying Destroyers:
HMS *Fearless*, HMS *Defender*, HMS *Acheron*, HMS *Ariel*, HMS *Attack*, HMS *Hydra*, HMS *Badger*, HMS *Lizard*, HMS *Goshawk*, HMS *Lapwing*

The German Fleet at the Battle of Jutland

The High Seas Fleet

Commander-in-chief: Adm Reinhard Scheer (SMS Friedrich der Grosse)

Battleships
SMS *Friedrich der Grosse* (Fleet Flagship)

3rd Battle Squadron: RAdm Paul Behncke
5th Division: SMS *König* (flagship), SMS *Grosser Kurfürst*, SMS *Kronprinz*, SMS *Markgraf*
6th Division: SMS *Kaiser* (flagship), SMS *Prinzregent Luitpold*, SMS *Kaiserin*

1st Battle Squadron: VAdm Ehrhard Schmidt
1st Division: SMS *Ostfriesland* (flagship), SMS *Thüringen*, SMS *Helgoland*, SMS *Oldenburg*
2nd Division: SMS *Posen* (flagship), SMS *Rheinland*, SMS *Nassau*, SMS *Westfalen*

2nd Battle Squadron: RAdm Franz Mauve
3rd Division: SMS *Deutschland* (flagship), SMS *Hessen*, SMS *Pommern*
4th Division: SMS *Hannover* (flagship), SMS *Schlesien*, SMS *Schleswig-Holstein*

Accompanying Light Cruisers:
4th Scouting Group: Kom Ludwig von Reuter
SMS *Stettin* (flagship), SMS *München*, SMS *Frauenlob*, SMS *Stuttgart*, SMS *Hamburg*

Accompanying Torpedo Boats (Destroyers):
First Leader of Torpedo-Boats: Cdre Andreas Michelsen, SMS *Rostock* (flotilla leader)
1st Torpedo-Boat Flotilla: Lt Conrad Albrecht
SMS *G39* (flotilla leader), SMS *G40*, SMS *G38*, SMS *S32*
3rd Torpedo-Boat Flotilla: Lt Cdr Wilhelm Hollmann
SMS S53 (flotilla leader), SMS *V71*, SMS *V73*, SMS *G88*, SMS *V48*, SMS *S54*, SMS *G42*
5th Torpedo-Boat Flotilla: Lt Cdr Oskar Heinecke
SMS *G11* (flotilla leader), SMS *V2*, SMS *V4*, SMS *V6*, SMS *V1*, SMS *V3*, SMS *G8*, SMS *V5*, SMS *G7*, SMS *G9*, SMS *G10*
7th Torpedo-Boat Flotilla: Lt Cdr Gottlieb von Koch
SMS *S24* (flotilla leader), SMS *S15*, SMS *S17*, SMS *S20*, SMS *S16*, SMS *S18*, SMS *S19*, SMS *S23*, SMS *V189*

Battle Cruiser Fleet
VAdm Franz Hipper

Battle Cruisers
SMS *Lützow* (flagship), SMS *Derfflinger*, SMS *Seydlitz*, SMS *Moltke*, SMS *Von der Tann*

Accompanying Light Cruisers:
2nd Scouting Group: RAdm Friedrich Boedicker
SMS *Frankfurt* (flagship), SMS *Elbing*, SMS *Pillau*, SMS *Wiesbaden*

Accompanying Torpedo Boats (Destroyers):
Second Leader of Torpedo-Boats: Cdre Paul Heinrich, SMS *Regensburg* (flagship)

2nd Torpedo-Boat Flotilla: Cdr Heinrich Schuur
SMS *B98* (flotilla leader), SMS *G101*, SMS *G102*, SMS *B112*, SMS *B97*, SMS *B109*, SMS *B110*, SMS *B111*, SMS *G103*, SMS *G104*
6th Torpedo-Boat Flotilla: Lt Cdr Max Schultz
SMS *G41* (flotilla leader), SMS *V44*, SMS *G87*, SMS *G86*, SMS *V69*, SMS *V45*, SMS *V46*, SMS *S50*, SMS *G37*
9th Torpedo-Boat Flotilla: Lt Cdr Herbert Goehle
SMS *V28* (flotilla leader), SMS *V27*, SMS *V26*, SMS *S36*, SMS *S51*, SMS *S52*, SMS *V30*, SMS *S34*, SMS *S33*, SMS *V29*, SMS *S35*

Glossary

abaft – towards the rear, or stern, of a boat.

cable – a sea term used for measuring distances less than a mile; i.e. 600 feet, one tenth of a nautical mile.

dhobey – term for cleaning laundry, coming from the Indian word 'dhob', meaning 'washing'.

dreadnought – a class of ship named after HMS *Dreadnought*, the first battleship of this type, in which most of the firepower is concentrated in guns of the same calibre.

fathom – a unit of length equal to 6 feet (1.8 m), used in reference to depth of water.

fighting top – a gun platform or site for a range-finder on the lower masts of a ship.

gunlayer – the person who controls the angle of elevation of a gun.

knot – a measure of speed reckoned in nautical miles per hour (although 'per hour' as such is never used in the Navy).

libertymen – Men excused duty and taking leave.

nautical mile – estimated at approximately 6,000 feet (1853m).

overs – projectiles striking or falling near a vessel that have passed over, or fallen short of, their intended target.

stoker – naval rating who worked in the boiler room.

sweep – a term used to denote patrolling; no connection with minesweeping.

trimmers – people who worked on all coal handling tasks, from loading coal into a bunker to bringing coal to the boilers.

Within Log entries:

a/c – altered course

blrs – boilers

co – course

fms – fathoms

red - reduce speed

s/c – set course

z.z – zigzag

Sources of Reference

Primary Sources:

Wirral Archives Service, Birkenhead:

'Launching Particulars' HMS *Caroline,* HMS *Constance,* pp 45/46, BAE Systems Ltd, formerly Cammell Laird

'Launching Arrangement', No. 5 Slipway Tranmere, drawings, BAE Systems Ltd, formerly Cammell Laird

'List of Vessels Built Since 1890 ', BAE Systems Ltd, formerly Cammell Laird

Trial Trip Reports, No.6 turbines, p137, BAE Systems Ltd, formerly Cammell Laird

National Archives, Kew:

S21b (Revised 1911) Deck log HMS *Caroline,* ADM53/37140

HR Crooke, Captain RN, personal service record

AP Smith Certificate of Service, ADM188 SHA-SoZ 1173 (L1-L15000) Officers' stewards

Captains' letters, Reports of Proceedings, ADM1. ADM137 for World War 1

Personal Property of Josie Loosemore, Chiddingly, West Sussex, AP Smith's daughter:

Personal diaries: Albion Percy Smith, 1916, 1917

Secondary Sources:

Allison, RS *HMS Caroline*, Blackstaff, 1974

Bennet, Geoffrey *The Battle of Jutland,* David and Charles, 1980

Chalmers, William Scott *The Life and Letters of David Beatty,* Hodder and Stoughton, 1951

Corbett, Sir Julian S *Official History of the War, Naval Operations: Official History of the War* Vols 1, Vol 1 (maps), Vol 2 & 3

Etienne *A Naval Lieutenant 1914–1918*, Methuen, 1919

Fairbairn, Douglas *The Narrative of A Naval Nobody, 1907–1924*, Murray, 1929

Faucett, Lt Cdr RN and Cooper, Lt RN, *The Fighting at Jutland*, Naval Institute Press 2001 (new edition)

Hall, Cdr. RN *My Naval Life*, Faber 1935

Hashagen, Ernst *The Log of a U-Boat Commander, 1914–1918*, Putnam, 1931

Jellicoe, Admiral of the Fleet, Sir John Rushworth *The Grand Fleet*, Cassell, 1919

Keyes, Sir Roger *The Naval Memoirs of Admiral of the Fleet, Vol 2, 1916–1918*, Thornton Butterworth, Ltd, 1934

Kroschel, Gunter *Die Deutsche Flotte, 1848–1945, Geschichte des Deutschen Kriegsschiffbaus*, Lohse-Eissing, Wilhelmshaven, 1963

Liddle, PH *Sailors' War, 1914–18*, Blandford Press, 1985 (second edition)

Macintyre, Captain Donald, DSO, DSC *Jutland*, Easton Press, 1958

Manning, Captain TD and Walker, Cdr CF *British Warship Names*, Putnam 1959

Marder, AJ, *From the Dreadnought to Scapa Flow, Vol 1, The Road to War; Vol 2, The War Years to the Eve of Jutland; 1914–16, Vol 3, Jutland and After*, Seaforth Publishing May 1916–Dec 1916

Nauticus, 1914, Year Book of the Imperial German Navy

Naval Rating Handbooks

Newbolt, Henry *Submarine and Anti-Submarine*, 1918

Protheroe, Ernest *The British Navy Its Making and Its Meaning*, Routledge, 1914

Scarborough Maritime Heritage Centre, 2015 exhibition on The Bombardment, December 1914

Scheer, Reinhard, Admiral *Deutschlands Hochseeflotte im Weltkrieg*, A Scherl, Berlin 1920

Shipbuilder Monthly magazine, 1914/15

Tirpitz, Grand Admiral Alfred von Tirpitz *My Memoirs*, Vols 1& 2, 1919

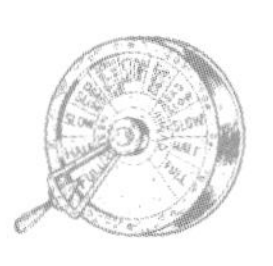

INDEX

Index of Vessels